NATIONAL ACADEMIES *Sciences Engineering Medicine*

NATIONAL ACADEMIES PRESS
Washington, DC

Facial Recognition Technology

Current Capabilities, Future Prospects, and Governance

Committee on Facial Recognition:
Current Capabilities, Future Prospects,
and Governance

Computer Science and Telecommunications Board

Division on Engineering and Physical Sciences

Committee on Science, Technology, and Law

Policy and Global Affairs Division

Committee on Law and Justice

Division of Behavioral and Social Sciences and Education

Consensus Study Report

NATIONAL ACADEMIES PRESS 500 Fifth Street, NW Washington, DC 20001

This activity was supported by the Department of Homeland Security (DHS) through contract number 70RSAT21G00000003/70RDAD21FR0000159 with the National Academy of Sciences and by the Federal Bureau of Investigation with the assistance of DHS. Any opinions, findings, conclusions, or recommendations expressed in this publication do not necessarily reflect the views of any organization or agency that provided support for the project.

International Standard Book Number-13: 978-0-309-71320-7
International Standard Book Number-10: 0-309-71320-X
Digital Object Identifier: https://doi.org/10.17226/27397
Library of Congress Control Number: 2024933216

This publication is available from the National Academies Press, 500 Fifth Street, NW, Keck 360, Washington, DC 20001; (800) 624-6242 or (202) 334-3313; http://www.nap.edu.

Copyright 2024 by the National Academy of Sciences. National Academies of Sciences, Engineering, and Medicine and National Academies Press and the graphical logos for each are all trademarks of the National Academy of Sciences. All rights reserved.

Printed in the United States of America.

Suggested citation: National Academies of Sciences, Engineering, and Medicine. 2024. *Facial Recognition Technology: Current Capabilities, Future Prospects, and Governance*. Washington, DC: The National Academies Press. https://doi.org/10.17226/27397.

The **National Academy of Sciences** was established in 1863 by an Act of Congress, signed by President Lincoln, as a private, nongovernmental institution to advise the nation on issues related to science and technology. Members are elected by their peers for outstanding contributions to research. Dr. Marcia McNutt is president.

The **National Academy of Engineering** was established in 1964 under the charter of the National Academy of Sciences to bring the practices of engineering to advising the nation. Members are elected by their peers for extraordinary contributions to engineering. Dr. John L. Anderson is president.

The **National Academy of Medicine** (formerly the Institute of Medicine) was established in 1970 under the charter of the National Academy of Sciences to advise the nation on medical and health issues. Members are elected by their peers for distinguished contributions to medicine and health. Dr. Victor J. Dzau is president.

The three Academies work together as the **National Academies of Sciences, Engineering, and Medicine** to provide independent, objective analysis and advice to the nation and conduct other activities to solve complex problems and inform public policy decisions. The National Academies also encourage education and research, recognize outstanding contributions to knowledge, and increase public understanding in matters of science, engineering, and medicine.

Learn more about the National Academies of Sciences, Engineering, and Medicine at **www.nationalacademies.org**.

Consensus Study Reports published by the National Academies of Sciences, Engineering, and Medicine document the evidence-based consensus on the study's statement of task by an authoring committee of experts. Reports typically include findings, conclusions, and recommendations based on information gathered by the committee and the committee's deliberations. Each report has been subjected to a rigorous and independent peer-review process and it represents the position of the National Academies on the statement of task.

Proceedings published by the National Academies of Sciences, Engineering, and Medicine chronicle the presentations and discussions at a workshop, symposium, or other event convened by the National Academies. The statements and opinions contained in proceedings are those of the participants and are not endorsed by other participants, the planning committee, or the National Academies.

Rapid Expert Consultations published by the National Academies of Sciences, Engineering, and Medicine are authored by subject-matter experts on narrowly focused topics that can be supported by a body of evidence. The discussions contained in rapid expert consultations are considered those of the authors and do not contain policy recommendations. Rapid expert consultations are reviewed by the institution before release.

For information about other products and activities of the National Academies, please visit www.nationalacademies.org/about/whatwedo.

**COMMITTEE ON FACIAL RECOGNITION:
CURRENT CAPABILITIES, FUTURE PROSPECTS, AND GOVERNANCE**

EDWARD W. FELTEN (NAE), Princeton University, *Co-Chair*
JENNIFER L. MNOOKIN, University of Wisconsin–Madison, *Co-Chair*
THOMAS D. ALBRIGHT (NAS), Salk Institute for Biological Studies
RICARDO BAEZA-YATES, Northeastern University
BOB BLAKLEY, Team8
PATRICK GROTHER, National Institute of Standards and Technology
MARVIN B. HAIMAN, Metropolitan Police Department, Washington, DC
AZIZ Z. HUQ, University of Chicago
ANIL K. JAIN (NAE), Michigan State University
ELIZABETH E. JOH, University of California, Davis
MICHAEL C. KING, Florida Institute of Technology
NICOL TURNER LEE, The Brookings Institution
IRA S. REESE, Global Security and Innovative Strategies
CYNTHIA RUDIN, Duke University

Study Staff

BRENDAN ROACH, Program Officer, Computer Science and Telecommunications Board (CSTB) (through December 31, 2023)
JON K. EISENBERG, Senior Board Director, CSTB
STEVEN KENDALL, Senior Program Officer, Committee on Science, Technology, and Law
GABRIELLE M. RISICA, Program Officer, CSTB
EMILY BACKES, Deputy Board Director, Committee on Law and Justice
SHENAE A. BRADLEY, Administrative Assistant, CSTB

NOTE: See Appendix D, Disclosure of Unavoidable Conflict of Interest.

COMPUTER SCIENCE AND TELECOMMUNICATIONS BOARD

LAURA HAAS (NAE), University of Massachusetts Amherst, *Chair*
DAVID DANKS, University of California, San Diego
CHARLES ISBELL, University of Wisconsin–Madison
ECE KAMAR, Microsoft Research
JAMES F. KUROSE (NAE), University of Massachusetts Amherst
DAVID LUEBKE, NVIDIA
JOHN L. MANFERDELLI, Independent Consultant, San Francisco, California, *Ex Officio*
DAWN C. MEYERRIECKS, MITRE Corporation
WILLIAM L. SCHERLIS, Carnegie Mellon University
HENNING SCHULZRINNE, Columbia University
NAMBIRAJAN SESHADRI (NAE), University of California, San Diego
KENNETH E. WASHINGTON, Medtronic

Staff

JON K. EISENBERG, Senior Board Director
SHENAE A. BRADLEY, Administrative Assistant
RENEE HAWKINS, Finance Business Partner
THƠ H. NGUYỄN, Senior Program Officer
GABRIELLE M. RISICA, Program Officer
BRENDAN ROACH, Program Officer
NNEKA UDEAGBALA, Associate Program Officer

COMMITTEE ON SCIENCE, TECHNOLOGY, AND LAW

MARTHA MINOW, Harvard Law School, *Co-Chair*
HAROLD VARMUS (NAS/NAM), Weill Cornell Medicine, *Co-Chair*
DAVID APATOFF, Arnold and Porter
ERWIN CHEMERINSKY, University of California, Berkeley, School of Law
ELLEN WRIGHT CLAYTON (NAM), Vanderbilt University Medical Center
JOHN S. COOKE, Federal Judicial Center, *Ex Officio*
JENNIFER EBERHARDT (NAS), Stanford University
KENNETH C. FRAZIER, Merck & Co., Inc.
CAROL W. GREIDER (NAS/NAM), University of California, Santa Cruz
STEVEN E. HYMAN (NAM), Harvard University
BARBARA McGAREY, Department of Health and Human Services (Retired)
ERNEST J. MONIZ, Massachusetts Institute of Technology
KIMANI PAUL-EMILE, Fordham University
K. SABEEL RAHMAN, Brooklyn Law School
NATALIE RAM, University of Maryland Francis King Carey School of Law
JULIE ROBINSON, U.S. District Court for the District of Kansas
PATTI B. SARIS, U.S. District Court for the District of Massachusetts
VICKI L. SATO, Denali Therapeutics and VIR Biotechnology, Inc.
BARBARA A. SCHAAL (NAS), Washington University in St. Louis
JOSHUA M. SHARFSTEIN (NAM), Johns Hopkins Bloomberg School of Public Health
CLIFFORD J. TABIN (NAS), Harvard Medical School

Staff

ANNE-MARIE MAZZA, Senior Director
STEVEN KENDALL, Senior Program Officer
RENEE DALY, Senior Program Assistant

COMMITTEE ON LAW AND JUSTICE

ROBERT D. CRUTCHFIELD, University of Washington, *Chair*
SALLY S. SIMPSON, University of Maryland, *Vice Chair*
ROD K. BRUNSON, University of Maryland
PREETI CHAUHAN, John Jay College of Criminal Justice
CYNTHIA LUM, George Mason University
JOHN M. MacDONALD, University of Pennsylvania
KAREN MATHIS, University of Colorado Boulder
THOEDORE A. McKEE, U.S. Court of Appeals for the Third Circuit
SAMUEL L. MYERS, JR., University of Minnesota
EMILY OWENS, University of California, Irvine
LAURIE O. ROBINSON, George Mason University, Consultant
ADDIE ROLNICK, University of Nevada, Las Vegas
WILLIAM J. SABOL, Georgia State University
VINCENT SCHIRALDI, Maryland Department of Juvenile Services
EMILY WANG, Yale School of Medicine

Staff

NATACHA BLAIN, Senior Board/Program Director
EMILY BACKES, Deputy Board Director
ABIGAIL ALLEN, Associate Program Officer
JULIE SCHUCK, Program Officer
STACEY SMIT, Program Coordinator

Reviewers

This Consensus Study Report was reviewed in draft form by individuals chosen for their diverse perspectives and technical expertise. The purpose of this independent review is to provide candid and critical comments that will assist the National Academies of Sciences, Engineering, and Medicine in making each published report as sound as possible and to ensure that it meets the institutional standards for quality, objectivity, evidence, and responsiveness to the study charge. The review comments and draft manuscript remain confidential to protect the integrity of the deliberative process.

We thank the following individuals for their review of this report:

JAMES A. BAKER, Harvard University Law School
R. ALTA CHARO (NAM), University of Wisconsin–Madison
RAMALINGAM CHELLAPPA (NAE), Johns Hopkins University
ELLEN WRIGHT CLAYTON (NAM), Vanderbilt University
STÉPHANE GENERIS, IDEMIA
SUSAN LANDAU, Tufts University
ALICE O'TOOLE, The University of Texas at Dallas
NIKKI POPE, NVIDIA Corporation

Although the reviewers listed above provided many constructive comments and suggestions, they were not asked to endorse the conclusions or recommendations of this report, nor did they see the final draft before its release. The review of this report was overseen by ROBERT F. SPROULL (NAE), University of Massachusetts Amherst, and JONATHAN D. MORENO (NAM), University of Pennsylvania Health System. They were responsible for making certain that an independent examination of this report was carried out in accordance with the standards of the National Academies and that all review comments were carefully considered. Responsibility for the final content rests entirely with the authoring committee and the National Academies.

Contents

PREFACE		xiii
SUMMARY		1
1	**INTRODUCTION**	19
	What Is Facial Recognition Technology?, 19	
	Expanding Scope and Scale, 20	
	Benefits and Concerns, 23	
	The Governance of Facial Recognition Technology, 26	
	About This Report, 30	
2	**FACIAL RECOGNITION TECHNOLOGY**	31
	Algorithms, 32	
	Image Acquisition, 42	
	Pose, Illumination, Expression, and Facial Aging Effects, 44	
	Accuracy, 48	
	Demographic Disparities, 55	
	Face Recognition Under Attack, 59	
	Human Roles and Capabilities, 61	
	Other Salient Attributes of Today's Commercial Facial Recognition Technology, 63	

3 USE CASES — 65
- Law Enforcement Investigation, 66
- Public Safety, 66
- In Lieu of Other Methods for Verifying Identity or Confirming Presence, 72
- Personal Device Access, 78
- Nonconsensual Commercial and Other Private Purposes, 79

4 EQUITY, PRIVACY, CIVIL LIBERTIES, HUMAN RIGHTS, AND GOVERNANCE — 81
- Equity, Race, and Facial Recognition Technology, 82
- Civil Liberties, Privacy, Human Rights, and Facial Recognition Technology, 87
- The Governance of Facial Recognition Technology, 91
- Facial Recognition Technology in Criminal Investigations and Trials, 102
- Addressing Wrongful Matches and Intrusive Deployment of Facial Recognition Technology, 104

5 CONCLUSIONS AND RECOMMENDATIONS — 107
- Technical Performance and Standards, 108
- Risk Management Framework, 111
- Applying the Framework to Real-World Use Cases, 117
- Use of Facial Recognition for Law Enforcement Investigations, 121
- Research and Development, 124
- Bias and Trustworthiness, 126
- Potential Executive Action and Legislation, 127

APPENDIXES
- A Statement of Task — 133
- B Presentations to the Committee — 135
- C Committee Member Biographical Information — 138
- D Disclosure of Unavoidable Conflict of Interest — 146

Preface

Facial recognition technology (FRT) is an increasingly prevalent tool for automated identification and identity verification. The use of FRT in a wide and growing variety of contexts has brought into increasing focus both the potential benefits of using FRT and concerns about impacts on equity, privacy, and civil liberties. In 2021, the Department of Homeland Security requested that the National Academies of Sciences, Engineering, and Medicine conduct a study that considers current capabilities, future possibilities, societal implications, and governance of FRT. The Federal Bureau of Investigation (FBI) joined as a formal sponsor of the study in March 2023.

The National Academies established the Committee on Facial Recognition: Current Capabilities, Future Prospects, and Governance to conduct this study (for biographical information, see Appendix C). The study addresses current use cases; explains how facial recognition technologies operate; and examines the legal, social, and ethical issues implicated by their use. The full statement of task for the committee is shown in Appendix A.

The committee met in person in July 2022 and February 2023 and met virtually 16 times to receive briefings from experts and stakeholders (for a list of presentations, see Appendix B), review relevant reports and technical literature, deliberate, and develop this report.

The committee would like to thank the Department of Homeland Security's Office of Biometric Identity Management (OBIM) and the FBI for their sponsorship of this study. It would also like to thank James L. Wayman, a member of the scientific staff in OBIM, and Richard W. Vorder Bruegge, a senior physical scientist at the FBI, who served as technical liaisons with the study. Last, the committee would like to acknowledge the assistance throughout the study of the following National Academies' staff: Brendan Roach, Steven Kendall, Gabrielle Risica, Shenae Bradley, Emily Backes, and Jon Eisenberg.

Edward W. Felten, *Co-Chair*
Jennifer L. Mnookin, *Co-Chair*
Committee on Facial Recognition:
   Current Capabilities, Future Prospects, and Governance

Summary

Facial recognition technology (FRT) is an increasingly prevalent tool for automated identification and identity verification of individuals. Its speed and accuracy have improved dramatically in the past decade. Its use speeds up identification tasks that would otherwise need to be performed manually in a slower or more labor-intensive way and, in many use cases, makes identification tasks practical that would be entirely infeasible without the use of these tools.

FRT measures the pairwise similarity of digital images of human faces to establish or verify identity. It uses machine learning models to extract facial features from an image, creating what is known as a *template*. It then compares these templates to compute a *similarity score*. In one-to-one comparison, the claimed identity of a single individual is verified by comparing the template of a captured *probe image* with an existing *reference image* (is this person who they say they are?). In one-to-many comparison, an individual is identified by comparing the template of a captured face image to the templates for many individuals contained in a database of reference images known as a *gallery* (what is the identity of the unknown person shown in this image?).

FRT accuracy is affected by image quality. Good quality is associated with cooperative capture in which the subject is voluntarily facing a good camera at close range with good lighting. Good lighting is especially important to give correct contrast in subjects with darker skin tones. Non-cooperative capture, in which subjects may not even realize that their image is being captured, such as images taken from security cameras, generally results in lower-quality images.

The attributes of FRT make it very useful in a number of identity verification and identification applications. These include the following:

- FRT enables the processing of large numbers of individuals quickly. For example, at international entry points, FRT allows arriving passengers to clear passport control faster.
- FRT makes it possible to identify high-risk individuals among large numbers of people entering a location without delaying others. FRT can, for example, be used to screen those entering a concert venue for individuals known to pose a threat to the performers.
- FRT can be a powerful aid for law enforcement in criminal and missing person investigations because it enables investigators to generate leads using images captured at a crime scene. A number of law enforcement agencies have reported successful use of FRT to generate otherwise unavailable leads.
- FRT can be especially convenient as a means of identity verification. For example, FRT allows a smartphone to be unlocked or a payment to be authorized without entering a passcode.

At the same time, FRT raises significant equity, privacy, and civil liberties concerns that merit attention by organizations that develop, deploy, and evaluate FRT—as well as government agencies, legislatures, state and federal courts, and civil society organizations (see the conclusions and Recommendations 3 and 4 in the following text). These concerns arise from such factors as FRT's low cost and ease of deployment, its ability to be used by inexperienced and inadequately trained operators, its potential for surveillance and covert use, the widespread availability of personal information that can be associated with a face image, and the observed differences in false negative (FN) and false positive (FP) match rates across phenotypes and demographic groups.

These are not just abstract or theoretical concerns:

- FRT can be a powerful tool for pervasive surveillance. Concerns about government, commercial, and private use are compounded by the potential to aggregate FRT matches over time to create a dossier of a person's activities, preferences, and associations—as has been the case in some authoritarian regimes.
- As FRT becomes more widespread and inexpensive, private individuals may have the means to use FRT against others in ways that raise troubling concerns about privacy and autonomy. Indeed, at least one online service already allows anyone to search for similar faces in a large gallery of images collected without explicit consent from the Web.

- There are significant concerns about adverse equity and privacy impacts in the largely unregulated commercial sphere and the implications of collecting massive databases of face images without consent or other safeguards.
- FRT has been implicated in at least six high-profile wrongful arrests of Black individuals. Although these incidents likely represent a small percentage of known arrests involving FRT, comprehensive data on the prevalence of FRT use, how often FRT is implicated in arrests and convictions, or the total number of wrongful arrests that have occurred on the basis of FRT use do not exist. Moreover, these incidents have occurred against a backdrop of deep and pervasive distrust by historically disadvantaged and other vulnerable populations of policing methods that have often included a variety of forensic, surveillance, and predictive technologies. The fact that all the reported wrongful arrests associated with the use of FRT have involved Black defendants exacerbates distrust of this technology. Concerningly, testing has demonstrated that FP match rates for Black individuals and members of some other demographic groups are relatively higher (albeit low in absolute terms) in FRT systems that are widely used in the United States.

Further compounding these concerns are many other potentially troubling uses—including uses that are technically feasible but not yet seen and uses that presently occur only outside the United States.

The National Academies of Sciences, Engineering, and Medicine undertook this study to assess current capabilities, future possibilities, societal implications, and governance of FRT. The study, sponsored by the Department of Homeland Security (DHS) and the Federal Bureau of Investigation, considers current use cases for FRT, explains how the technology works, and examines the legal, social, and ethical issues implicated by its use.

Deemed out of scope for this study are related computational techniques that classify a face image as a member of a given category, such as race, gender, or age, or to identify specific activities, behaviors, or characteristics of an individual not leading to an identification or verification and that are not normally considered face *recognition* technology.

CONCLUSIONS

FRT has matured into a powerful technology for identification and identity verification. Some uses offer convenience, efficiency, or enhanced safety, while others—including ones already deployed in the United States—are troubling and raise significant equity, privacy, and civil liberties concerns that have not been resolved by U.S. courts or legislatures.

Concerns about the use of FRT arise from two (non-exclusive) factors that require different analysis and merit different policy responses:

- *Concerns about poor performance of the technology*—for example, unacceptable FP or FN rates or unacceptable variation of these rates across demographic groups, especially in the case of poor-quality surveillance images.
- *Concerns about problematic use or misuse of the technology*—for example, technology with acceptable technical performance sometimes produces societally undesirable outcomes as a result of either inadequate procedures or training for operating, evaluating, or making decisions using FRT or the deliberate use of FRT to achieve a societally undesirable outcome, including uses not foreseen by FRT developers or vendors.

That is, some concerns about FRT can be addressed by improving the technology, while others require changes to procedures or training, restrictions on when or how FRT is used, or regulation of the conduct that FRT enables. Furthermore, some uses of FRT may well cause such concern that they should be not only regulated but prohibited.

Currently, with a few exceptions, such as new department-wide guidance issued by DHS in September 2023, the nation does not have authoritative guidance, regulations, or laws that adequately address these concerns broadly.

Much progress has been made in recent years to characterize, understand, and mitigate phenotypical disparities in the accuracy of FRT results. However, these performance differentials have not been entirely eliminated, even in the most accurate existing algorithms. FRT still performs less well for individuals with certain phenotypes, including those typically distinguished on the basis of race, ethnicity, or gender.

Tests show that FN rate differentials are extremely small, especially with the most accurate algorithms and if both the probe and reference images are of high quality, but can become significant if they are not. FN matches occur when the similarity score between two different images of the same person is low. Causes include changes in appearance and loss of detail from poor image contrast. FN match rates vary across algorithms and have been measured to be higher by as much as a factor of 3 in women,

Africans, and African Americans than in Whites. The algorithms that have the highest overall accuracy rates also generally have the lowest demographic variance. FN match rate disparities are highest in applications where the photographic conditions cannot be controlled; they are lower in circumstances with better photography and better comparison algorithms. The consequences of an FN match include a failure to identify the subject of an investigation or the need for an individual to identify themselves in another way such as by presenting identity documents. Rate disparities mean, for example, that the burden of presenting identification or facing additional questioning currently falls disproportionately on some groups of individuals—including groups that have been historically disadvantaged and marginalized. Although this additional time and inconvenience may be seemingly small in a single instance, the aggregate impacts to individuals who repeatedly encounter it and to groups disproportionately affected can be large.

Tests also show that for identify verification (one-to-one comparison) algorithms, the FP match rates for certain demographic groups when using even the best-performing facial recognition algorithms designed in Western countries and trained mostly on White faces are relatively higher (albeit quite low in absolute terms), even if both the probe and the reference images are of high quality. Demographic differentials present in verification algorithms are usually but not always present in identification (one-to-many comparison) algorithms.

FP matches occur when the similarity score between images of two different people is high. (Instances of FP matches can thus be reduced with a higher similarity threshold.) Higher FP match rates are seen with women, older subjects, and—for even the best-performing FRT algorithms designed in Western countries and trained mostly on White faces—individuals of East Asian, South Asian, and African descent. With current algorithms, FP match rate differences occur even when the images are of very high quality and can vary across demographic groups markedly and contrary to the intent of the developer. However, some Chinese-developed algorithms have the lowest FP rates for East Asian subjects, suggesting that the makeup of faces in the training database, rather than some inherent aspect of FRT, contributes to these results. FP match rate disparities can therefore likely be reduced by using more diverse data to train models used to create templates from facial images or by model training with a loss function that more evenly clusters but separates demographic groups. The applications most affected by FP match rate differentials are those using large galleries and where most searches are for individuals who are not present in the gallery. FP rate disparities will mean that members of some groups bear an unequal burden of, for example, being falsely identified as the target of an investigation.

A final concern with FPs is that as the size of reference galleries or the rate of queries increases, the possibility of an FP match grows, because there are more potential templates that can return a high similarity score to a probe face. Some facial recognition algorithms, however, adjust similarity scores in an attempt to make the FP match rate independent of the gallery size.

With respect to the need for regulation of FRT, the committee concluded that an outright ban on all FRT under any condition is not practically achievable, may not necessarily be desirable to all, and is in any event an implausible policy, but restrictions or other regulations are appropriate for particular use cases and contexts.

At the same time, the committee observes that because FRT has the potential for mass surveillance of the population, courts and legislatures will need to consider the implications for constitutional protections related to surveillance, such as due process and search and seizure thresholds and free speech and assembly rights.

In grappling with these issues, courts and legislatures will have to consider such factors as who uses FRT, where it is used, what it is being used for, under what circumstances it is appropriate to use FRT-derived information provided by third parties, whether FRT use is based on individualized suspicion, intended and unintended consequences, and susceptibility to abuse.

As governments and other institutions take affirmative steps through both law and policy to ensure the responsible use of FRT, they will need to consider the views of government oversight bodies, civil society organizations, and affected communities to develop appropriate safeguards.

Study committee members all agreed that some use cases of FRT should be permissible, that some use cases should be allowed only with significant limits or regulation, and that others likely should be altogether prohibited. But committee members did not reach a fully shared consensus on precisely which use cases should be permitted and how permitted uses should be regulated or otherwise governed, reflecting the complexity of the issues raised; their individual assessments of the risks, benefits, and trade-offs; and their perspectives on the underlying values. However, the committee is in full agreement with the following recommendations.

MITIGATING POTENTIAL HARMS AND LAYING THE GROUNDWORK FOR MORE COMPREHENSIVE ACTION

RECOMMENDATION 1: The federal government should take prompt action along the lines of Recommendations 1-1 through 1-6 to mitigate against potential harms of facial recognition technology and lay the groundwork for more comprehensive action.

RECOMMENDATION 1-1: The National Institute of Standards and Technology should sustain a vigorous program of facial recognition technology testing and evaluation to drive continued improvements in accuracy and reduction in demographic biases.

Testing and standards are a valuable tool for driving performance improvements and establishing appropriate testing protocols and performance benchmarks, providing a firmer basis for justified public confidence, for example, by establishing an agreed-on baseline of performance that a technology must meet before it is deployed. The National Institute of Standards and Technology's (NIST's) Facial Recognition Technology Evaluation has proven to be a valuable tool for assessing and thereby propelling advances in FRT performance, including by increasing accuracy and reducing demographic differentials.

RECOMMENDATION 1-2: The federal government, together with national and international standards organizations (or an industry consortium with robust government oversight), should establish
  a. Industry-wide standards for evaluating and reporting on the performance—including accuracy and demographic variation—of facial recognition technology products for private or public use.
  b. A tiered set of profiles that define the minimum quality for probe and reference images, acceptable overall false positive and false negative rates, and acceptable thresholds for accuracy variation across different phenotypes for applications of different sensitivity levels. It would be up to those creating guidance, standards, or regulations to select the appropriate profile for the application in question.
  c. Methods for evaluating false positive match rates for probe images captured by closed-circuit television or other low-resolution cameras (which have been implicated in erroneous arrests of several Black individuals).
  d. Process standards in such areas as data security and quality control.

NIST would be a logical home for such activities within the federal government, given its role in measurement and standards generally and FRT evaluation specifically.

The following two subrecommendations apply to law enforcement's use of FRT to identify suspects in criminal investigations.

RECOMMENDATION 1-3: The Department of Justice and the Department of Homeland Security should establish a multi-disciplinary and multi-stakeholder working group on facial recognition technology (FRT) to develop and periodically review standards for reasonable and equitable use, as well as other needed guidelines and requirements for the responsible use of FRT by federal, state, and local law enforcement. That body, which should include members from law enforcement, law enforcement associations, advocacy and other civil society groups, technical experts, and legal scholars, should be charged with developing

a. Standards for appropriate, equitable, and fair use of FRT by law enforcement.
b. Minimum technical requirements for FRT procured by law enforcement agencies and a process for periodically reevaluating and updating such standards.
c. Standards for minimum image quality for probe images, below which an image should not be submitted to an FRT system because of low confidence in any ensuing match. Such standards would need to take into account such factors as the type of investigation (including the severity of the crime and whether other evidence is available) and the resources available to the agency undertaking the investigation.
d. Guidance for whether FRT systems should (1) provide additional information about confidence levels for candidates or (2) present only an unranked list of candidates above an established minimum similarity score.
e. Requirements for the training and certification of law enforcement officers and staff and certification of law enforcement agencies using FRT as well as requirements for documentation and auditing. An appropriate body to audit this training and certification should also be identified.
f. Policies and procedures to address law enforcement failures to adhere to procedures or failure to attain appropriate certification.
g. Mechanisms for redress by individuals harmed by FRT misuse or abuse, including both damages or other remedies for individuals and mechanisms to correct systematic errors.
h. Policies for the use of FRT for real-time police surveillance of public areas so as to not infringe on the right of assembly or to discourage legitimate political discourse in public places, at political

gatherings, and in places where personally sensitive information can be gathered, such as schools, places of worship, and health care facilities.

i. Retention and auditing requirements for search queries and results to allow for proper oversight of FRT use.

j. Guidelines for public consultation and community oversight of law enforcement FRT.

k. Guidelines and best practices for assessing public perceptions of legitimacy and trust in law enforcement use of FRT.

l. Policies and standardized procedures for reporting of statistics on the use of FRT in law enforcement, such as the number of searches and the number of arrests resulting from the use of FRT, to ensure greater transparency.

RECOMMENDATION 1-4: Federal grants and other types of support for state and local law enforcement use of facial recognition technology (FRT) should require that recipients adhere to the following technical, procedural, and disclosure requirements:

a. Provide verified results with respect to accuracy and performance across demographics from the National Institute of Standards and Technology's Facial Recognition Technology Evaluation or similar government-validated third-party test.

b. Comply with the industry standards called for in Recommendation 1-2—or comply with future certification requirements, where certification would be granted on the basis of an independent third-party audit.

c. Use FRT systems that present only candidates who meet a minimum similarity threshold (and return zero matches if no candidates meet the threshold) rather than returning a fixed-length candidate list or "most-probable candidate" list when the output of an FRT system is being used for further investigation.

d. Adopt minimum standards for the quality of both probe and reference gallery images.

e. Use FRT systems only with a "human-in-the-loop" and not for automated detection of offenses, including issuing citations.

f. Limit the use of FRT to being one component of developing investigative leads. Given current technological capabilities and limitations, in light of present variations in training and protocols, and to ensure accountability and adherence with legal standards, FRT should be only part of a multi-factor basis for an arrest or investigation, in line with current fact-sensitive determinations of probable cause and reasonable suspicion.

g. Restrict operation of FRT systems to law enforcement organizations that have sufficient resources to properly deploy, operate, manage, and oversee them (an adequate certification requirement would presumably ensure that such resources were in place).

h. Adopt policies to disclose to criminal suspects, their lawyers, and judges on a timely basis the role played by FRT in law enforcement procedural actions, such as lead identification, investigative detention, establishing probable cause, or arrest.

i. Disclose to suspects and their lawyers, on arrest and in any subsequent charging document, that FRT was used as an element of the investigation that led to the arrest and specify which FRT product was used.

j. Publicly report on a regular basis de-identified data about arrests that involve the use of matches reported by FRT. The reports should identify the FRT system used, describe the conditions of use, and provide statistics on the occurrences of positive matches, false positive matches, and non-matches.

k. Publicly report cumulatively on any instances where arrests made partly on the basis of FRT are found to have been erroneous.

l. Conduct periodic independent audits of the technical optimality of an FRT system and the skills of its users, determining whether its use is indeed cost-justified.

Even if not subject to federal grant conditions, state and local agencies should adopt these standards.

RECOMMENDATION 1-5: The federal government should establish a program to develop and refine a risk management framework to help organizations identify and mitigate the risks of proposed facial recognition technology applications with regard to performance, equity, privacy, civil liberties, and effective governance.

Risk management frameworks are a valuable tool for identifying and managing sociotechnical risks, defining appropriate measures to protect privacy, ensuring transparency and effective human oversight, and identifying and mitigating concerns around bias and equity. A risk management framework could also form the basis for future mandatory disclosure laws or regulations. Current examples of federally defined risk management frameworks include NIST's Cybersecurity Framework and NIST's Artificial Intelligence Risk Management Framework. NIST would be a logical organization to be charged with developing this framework given its prominent role in FRT testing and evaluation as well as in developing risk management frameworks for other technologies.

Some issues that might be addressed by the framework are

- *Technical performance*—including accuracy and differential performance across standardized demographic groups, quality standards for probe and reference images, and adequate indication of the confidence of reported matches.
- *Equity*—including the extent to which there are statistically and materially significantly different probabilities of error for different demographic groups, the extent to which these are attributable to technical characteristics or other factors (e.g., the manner in which an FRT tool is used), and the parity of use among different populations.
- *Privacy*—including privacy protection for faces used in training the template extraction model, whether use of FRT significantly increases the scope or scale of the identification being performed, or other adverse privacy impact.
- *Data collection, disclosure, use, and retention policies for both subject and reference images and templates*—including data retention policies to limit, for example, inappropriate use of probe images for searches beyond pre-defined operational needs.
- *Data security and integrity*—including adequately protecting information in FRT training data sets and reference databases from exfiltration and misuse.
- *Civil liberties*—including whether FRT is being used to control access to a public benefit or service and whether the use of FRT will have a reasonably foreseeable negative impact on the exercise of civil rights, such as free speech or assembly, whether by individuals or groups.
- *Governance*—including whether there is an important public interest or legitimate business purpose; who decides whether and how to deploy FRT, and who assumes the risks and accrues the benefits of its use; consultation with the public at large or with affected groups, and meaningful consideration of results; and appropriate safeguards, oversight, and quality assurance.
- *Disclosure*—including meaningful public disclosure about where, when, and for what purpose the system is used. Transparency and standardized

reporting become more important in use cases where there are greater consequences for mistakes and errors.
- *Consent*—including whether consent is opt-in or opt-out and whether consent is meaningful and uncoerced, and in the case of mandatory use, whether the justification is clear and compelling.
- *Training*—including what sort of capabilities or competencies the operator of an FRT system, and those using its output, need to demonstrate and whether the training or certification regimes meet the needs of the system usage.
- *Human-in-the-loop*—including whether there is an individual responsible for all significant decisions made on the basis of an FRT match.
- *Accountability*—including who is responsible for addressing systematic technical issues with an FRT system, the manner in which it is used, ethical and societal concerns that arise from the social environment in which it is used, and whether and how frequently audits are conducted.
- *Adverse impacts and their distribution*—including the potential adverse impacts of an FP or FN match in the proposed use, identifying who bears the consequences of those impacts, and indicating whether costs are borne primarily by the individual subject or the operator of the technology.
- *Recourse*—including whether recourse mechanisms provide redress proportional to potential consequences, whether they are available to individuals who will experience adverse outcomes, and whether the organization has a mechanism for receiving complaints.

Note that some of the issues listed here cut across most, if not all, FRT use cases, while others are specific to particular use cases.

RECOMMENDATION 1-6: The federal government should support research to improve the accuracy and minimize demographic biases and to further explore the sociotechnical dimensions of current and potential facial recognition technology uses.

Public research organizations, such as NIST, already undertake important work in setting benchmarks and evaluating the performance of FRT systems. Additional government support could help NIST answer important questions on the performance of FRT systems in non-cooperative settings, how to improve data sets to both preserve privacy and promote equity in the performance of FRT tools, and how best to continue recent work on characterizing, understanding, and mitigating phenotypical disparities. To understand better how to responsibly deploy FRT while protecting equity, fairness, and privacy, NIST, DHS's Maryland Test Facility, or a similarly well-suited institution should conduct research on

- The accuracy of FRT systems in a variety of non-optimal settings, including non-optimal facial angle, focus, illumination, and image resolution.
- The development of representative training data sets for template extraction and other methods that developers can safely apply to existing data sets and models to adjust for demographic mismatches between a given data set and the public.
- The performance of FRT with very large galleries (i.e., tens or hundreds of millions of entries), to better understand the impacts of FP and FN match rates as the size of galleries used continues to grow.

To advance the science of FRT and to better understand the sociotechnical implications of FRT use, the National Science Foundation or a similar research sponsor should support research on

- Developing privacy-preserving methods to prevent malicious actors from reverse-engineering face images from stored templates.
- Mitigating FP match rate variance across diverse populations, and building better understanding of the levels at which residual disparities will not significantly affect real-world performance.
- Developing approaches that can reduce demographic and phenotypical disparities in accuracy.
- Developing accurate and fast methods for directly matching an encrypted probe image template to an encrypted template or gallery—for example, using fully homomorphic encryption.
- Developing robust methods to detect face images that have been deliberately altered by either physical means such as masks, makeup, and other types of alteration or by digital means such as computer-generated images.
- Determining whether FRT use deters people from using public services, particularly members of marginalized communities.
- Determining how FRT is deployed in non-cooperative settings, public reaction to this deployment, and its impact on privacy.
- Determining how FRT may be used in the near future by individuals for abusive purposes, including domestic violence, harassment, political opposition research, etc.

- Determining how private actors might use FRT in ways that mimic government uses, such as homeowners who deploy FRT for private security reasons.
- Researching future uses of FRT, and their potential impacts on various subgroups of individuals.

FOSTERING TRUST AND MITIGATING BIAS AND OTHER RISKS

RECOMMENDATION 2: Developers and deployers of facial recognition technology should employ a risk management framework and take steps to identify and mitigate bias and cultivate greater community trust.

RECOMMENDATION 2-1: Organizations deploying facial recognition technology (FRT) should adopt and implement a risk management framework addressing performance, equity, privacy, civil liberties, and effective governance to assist with decision making about appropriate use of FRT.

Until the recommended risk management framework is developed, the issues listed in Recommendation 1-5 may serve as a useful point of departure. Future standards documents may also provide relevant guidance.

RECOMMENDATION 2-2: Institutions developing or deploying facial recognition technology should take steps to identify and mitigate bias and cultivate greater community trust—with particular attention to minority and other historically disadvantaged communities. These should include

a. **Adopting more inclusive design, research, and development practices.**
b. **Creating decision-making processes and governance structures that ensure greater community involvement.**
c. **Engaging with communities to help individuals understand the technology's capabilities, limitations, and risks.**
d. **Collecting data on false positive and false negative match rates in order to detect and mitigate higher rates found to be associated with particular demographic groups.**

Such practices are imperative to help address mistrust about bias in FRT's technological underpinnings and to respond to broader mistrust, especially in communities of color, about the role of technology in law enforcement and similar contexts.

ENACTING MORE COMPREHENSIVE SAFEGUARDS

RECOMMENDATION 3: The Executive Office of the President should consider issuing an executive order on the development of guidelines for the appropriate use of facial recognition technology by federal departments and agencies and addressing equity concerns and the protection of privacy and civil liberties.

RECOMMENDATION 4: New legislation should be considered to address equity, privacy, and civil liberties concerns raised by facial recognition technology, to limit harms to individual rights by both private and public actors, and to protect against its misuse.

Legislation should consider the following:

a. *Limitations on the storing of face images and templates.* Legislation could, for example, prohibit the storing of face images or templates in a gallery unless the gallery will be used for a specifically allowed purpose. Inclusion in a gallery might, for example, be prohibited except under the following conditions:
   - *For prescribed government functions,* such as at the border or at international arrival and departure points to identify persons entering and leaving the country, using photos from government databases.
   - *Where there is explicit consent for a specific purpose,* such as a person setting up a new smartphone consenting to using FRT to unlock the phone or a person explicitly consenting to an airline's use of their passport photo to enable the person to check in and board flights using FRT.
   - *Where there are threats to life and physical safety,* such as by a performance venue to scan for specific individuals who have been reported by police as posing a threat to the life or physical safety of a performer or by a shelter for abuse victims to scan people arriving at the facility to find individuals subject to restraining orders prohibiting their interaction with residents of the shelter.

An additional set of issues with respect to inclusion in galleries relates to collection and use of images gathered from websites and social media platforms—both whether it is appropriate to use these without consent or knowledge as well as the implications of including low-quality or synthetic images collected in this manner. Under current law, the fact that a gallery was created by harvesting facial images from the Web in violation of platforms' terms of service does not create a barrier to the instrument's usage. Of course, Congress, a state legislature, or even a policing authority could promulgate a new rule barring the use of FRT applications developed without the benefit of consent from those whose data is used for training purposes.

Precisely which uses are or are not allowed merits careful consideration by legislators and the public at large. The risk management framework discussed earlier may provide a useful tool for considering these questions.

b. *Specific uses of concern.* Such uses might, for example, include the following:
- Commercial practices that implicate privacy (through either broader privacy legislation addressing FRT risks or an FRT-specific federal privacy law);
- Harassment or blackmail;
- Unwarranted exclusion from public or quasi-public places;
- Especially sensitive government FRT uses (e.g., pertaining to law enforcement or access to public benefits or federally subsidized housing);
- Public and private uses that tend to chill the exercise of political and civil liberties—both intentional and from the emergent properties of use at scale; and
- Mass surveillance or individual surveillance other than that properly authorized for law enforcement or national security purposes.

c. *User training.* In applications where the operator or other user is expected to apply judgment or discretion in when or how to use FRT systems or in interpreting their results, and where a false match may result in significant consequences for an individual, legislation could require training for the operators and decision makers. A notable example of this type of application is law enforcement investigations. By contrast, there are applications where the fallback in case of a failure is simply to inspect a government-issued identity document; training may be less critical for such use cases.

d. *Certification.* Legislation could require certification of operators and other users and/or certification of organizations that operate FRT systems for applications where technical or procedural errors can significantly harm subjects, notably in law enforcement.

* * *

FRT is a powerful tool with profound societal implications. It will be critically important to adopt a considered approach to its governance and future development.

1

Introduction

The use of facial recognition technology (FRT) in a wide and growing variety of contexts has brought into increasing focus both the potential benefits of using FRT and concerns about impacts on equity, privacy, and civil liberties.

WHAT IS FACIAL RECOGNITION TECHNOLOGY?

Facial recognition connects an image of a face to an identity or connects an image of a face to a database entry supporting identification or association with a prior event. Manual comparison of images of faces by humans is a long-standing practice that is slow, has less than perfect accuracy, and is subject to human biases.[1,2,3] By contrast, computer performance of facial recognition tasks is extremely quick and, in many cases, more accurate than human face comparisons.

Modern FRT uses an artificial intelligence (AI) model, typically deep convolutional neural networks, to extract facial features in each image, and then compares the extracted features (not the images themselves) between two images. It can either verify identity by matching a subject image to a record of a single individual (one-to-one

---

[1] N. Whitehead, 2014, "Face Recognition Algorithm Finally Beats Humans," *Science,* April 23, https://www.science.org/content/article/face-recognition-algorithm-finally-beats-humans.

[2] P.J. Phillips and A.J. O'Toole, 2014, "Comparison of Human and Computer Performance Across Face Recognition Experiments," *Image and Vision Computing* 32(1):74–85.

[3] A.J. O'Toole, P.J. Phillips, F. Jiang, J. Ayyad, N. Penard, and H. Abdi, 2006, *Face Recognition Algorithms Surpass Humans,* Washington, DC: National Institute of Standards and Technology, https://www.nist.gov/system/files/documents/2021/05/12/frgc_face_recognition_algorithms_surpasshumans.pdf.

matching) or identify an individual by matching the image to a record of an individual in a reference database (one-to-many matching). FRT can

- Associate a face with an identity to allow the person to later verify an identity claim. (Are you who you claim to be?)
- Associate a face with a database entry containing identification data. (Is this person known to us?)
- Match a face with an event or circumstance. (Has this person been seen before?)
- Provide evidence for who an observed person is not. (This is not the person we are looking for.)

Given these abilities, FRT is often used in forensic applications, helping to establish the identity of an unknown perpetrator using still images or video footage much in the way that fingerprint analysis can establish identity using latent prints. However, the applications of FRT extend well beyond forensic uses.

The term "face recognition" is sometimes confusingly misapplied to algorithms that estimate some property of an individual based on analysis of a face image. These include estimation tasks (e.g., how old is this person?); classification tasks (e.g., what sex is this person? does the person smoke?); and a multitude of other aspirational purposes such as emotion or mood determination or disease detection. It is important to make this distinction: Whereas face classification algorithms analyze one image, face recognition algorithms operate by comparing two images, entailing use of entirely different machinery. Moreover, some classification tasks used in the past, such as criminality or sexuality determination, have been debunked and are understood to be unethical pseudoscience. Such non-recognition face analysis capabilities are not considered in this report.

EXPANDING SCOPE AND SCALE

The technical development of FRT goes back more than 50 years but has accelerated greatly in the past decade with the adoption of deep convolutional neural network techniques from AI, the training of models that extract facial features from large numbers of face images, the curation of increasingly large data sets of facial images (often acquired without the consent of those whose faces are used), and experience gained from industrial adoption and deployment. The term FRT references a large number and variety of face recognition systems that are produced by an array of vendors, each of which uses its own algorithms, data sets used to train the models, and data sets used for comparison.

The acceleration in development, which continues today, has led to deployment in many different applications. FRT is now widely used to unlock smartphones and other personal devices. It is also used increasingly in law enforcement investigations, at international borders, and in airports—as well as in a variety of other government and commercial applications.

Government agencies amass large databases of facial imagery in the process of issuing identity documents such as drivers' licenses and passports and through the collection of mugshots as part of arrest procedures. Private entities also build databases using images collected from the Internet or on their premises. The boundaries between public and private FRT databases are fluid; law enforcement agencies, for example, regularly make use of databases created by private entities. As a consequence, they can assemble databases of face images and apply FRT to verify identity and make decisions regarding access. Government and commercial databases together make it possible in theory for government agencies to identify a large portion of the U.S. population using FRT.[4]

Although the market for FRT is relatively young and fragmented across a number of smaller vendors, it is growing rapidly. A 2020 industry survey estimated that the market for FRT was about $4 billion, with an anticipated annual growth rate of about 15 percent over the subsequent decade. An industry estimate suggests that by 2030, the global market for FRT will be worth nearly $17 billion.[5] The technology has become particularly prevalent in law enforcement contexts, with 20 out of 42 federal law enforcement agencies using the technology, according to a 2021 Government Accountability Office report.[6]

In addition to the proliferation of the use of FRT in law enforcement, such systems are increasingly being used at airports and other travel hubs. The Transportation Security Administration (TSA) has now expanded a pilot program to use FRT to verify traveler identity at security checkpoints in 25 airports across the United States.[7] Meanwhile, Customs and Border Protection has deployed FRT to track travelers exiting the country at 32 airports in the United States, and to track travelers entering the country at every international airport in the country.[8]

Coupled with the expansion of face recognition software and enabling the increasing efficacy of these technologies is the growth of large databases of face images.

---

[4] It is difficult to estimate the precise number of unique database entries available to government agencies, because not all sources are available to or used by any given agency.

[5] Allied Market Research, 2023, "Facial Recognition Market," https://www.alliedmarketresearch.com/facial-recognition-market.

[6] Government Accountability Office (GAO), 2021, "Facial Recognition Technology: Federal Law Enforcement Agencies Should Have Better Awareness of Systems Used by Employees," https://www.gao.gov/products/gao-21-105309.

[7] K.V. Cleave, 2023, "TSA Expands Controversial Facial Recognition Program," *CBS News*, June 5, https://www.cbsnews.com/news/tsa-facial-recognition-program-airports-expands.

[8] GAO, 2022, "Facial Recognition Technology: CBP Traveler Identity Verification and Efforts to Address Privacy Issues," https://www.gao.gov/products/gao-22-106154.

Governments routinely collect large numbers of face images for administrative purposes such as drivers' licenses and mugshots, and can, where legally authorized, use these images to create very large face recognition reference databases with tens of millions of faces. Meanwhile, private companies have also created very large face reference databases—for example, by collecting billions of images from social media and other websites. Most notably, Clearview AI claims to have collected a database of 30 billion face images (presumably including duplicates and synthetic face images) by collecting images from social media sites.[9] In the absence of robust privacy protections for face images and other biometric information, government agencies as well as private-sector organizations can use FRT systems that search against these resulting reference databases.

The application of FRT to numerous and growing sources of camera footage is another contributor to the potential scale of use. Conventional security camera footage has long been used by many businesses and in police investigations. With the falling cost of high-quality cameras, networking, and storage, the use of private cameras, such as doorbell-type cameras and cell phone cameras, has increased dramatically in recent years and can provide capture of additional images that can be used for FRT for investigative purposes. Furthermore, many cities have incentivized private security camera ownership,[10] with some offering rebates to cover costs of installing surveillance cameras for the purpose of deterring crime and facilitating criminal investigations. Private video footage is now commonly accessible by law enforcement investigators. As a result, FRT acts as both part of new technologies provided to individuals and law enforcement as well as a technology that can be retrofitted onto existing surveillance structures.

Indeed, there are numerous categories of current or potential use for FRTs. They range from somewhat innocuous uses that pose relatively modest equity, privacy, or civil liberties issues to potential uses that raise significant ethical and legal questions. Although not necessarily comprehensive, this report identifies several categories intended to illustrate this range of current and potential use:

- Law enforcement investigation of a specific lead or criminal act and prosecution.
- Preventive public safety or national security—such as screening for specific individuals known to pose a high risk at a venue or identifying known shoplifters at a retail store.

---

[9] K. Tangalakis-Lippert, 2023, "Clearview AI Scraped 30 Billion Images from Facebook and Other Social Media Sites and Gave Them to Cops: It Puts Everyone into a 'Perpetual Police Line-Up,'" *Business Insider*, updated April 3, https://www.businessinsider.com/clearview-scraped-30-billion-images-facebook-police-facial-recogntion-database-2023-4.

[10] Office of Victim Service and Justice Grants, "The Private Security Camera Rebate Program," Washington, DC Government, https://ovsjg.dc.gov/page/private-security-camera-rebate-program, accessed November 16, 2023.

- In lieu of other methods for verifying identity or presence—such as at international borders and entry/exit points, and to control employee access to workplaces.
- Personal device access.
- Non-opt-in, for commercial and other private purposes—such as retail stores identifying high-value customers.

Chapter 3 describes these categories in more detail and provides illustrative vignettes for each.

BENEFITS AND CONCERNS

FRT is far from the first technology to be used for identification or whose introduction has raised privacy concerns or led to challenges over potentially yielding inequitable outcomes. A number of identification technologies are used in forensic and non-forensic applications, including fingerprint, handprint, iris, and DNA comparison. Video captured by surveillance cameras has long been reviewed by humans to identify potential criminal perpetrators. The location of cell phones, carried by most of the population, can be tracked, and license plate readers can be used to track the movements of motor vehicles. Some of the benefits and concerns raised by FRT are familiar from the earlier technologies, while others are new or heightened by virtue of FRT's characteristics.

FRT is inexpensive, scalable, and contactless, and it can operate remotely in a covert manner. It allows existing identification tasks to be performed more efficiently than if done manually and enables new identification tasks that would otherwise be impractical. In particular,

- FRT enables the processing of large numbers of individuals quickly. For example, at international entry points, FRT allows arriving passengers to clear passport control very rapidly.
- FRT makes it possible to identify high-risk individuals among large numbers of people entering a location without delaying others. FRT can, for example, be used to screen those entering a concert venue for individuals known to pose a threat to the performers.
- FRT can be a powerful aid for law enforcement in criminal and missing person investigations because it enables investigators to generate leads using images captured at a crime scene. A number of law enforcement agencies have reported successful use of FRT to generate otherwise unavailable leads.

- FRT can be especially convenient as a means of identity verification. For example, FRT allows a smartphone to be unlocked or a payment to be authorized without entering a passcode.

However, there are significant concerns about FRT and the societal implications of its use. These include the following:

- *Significant demographic disparities in the performance of FRTs.* A number of studies have identified phenotypical disparities (see Chapter 2) and suggest the need for a better understanding of potential biases and disparities in face recognition systems.[11,12,13]
    - Drivers of demographics-related performance variation include photography not well adapted to dark skin tones and the under-representation of demographic groups in the training data used to create the models used to extract facial features. Considerable work has been done to understand sources of bias and demographic disparities better and to design robust models and to rigorously evaluate them on large-scale data sets to drive performance improvements.
    - Early FRT systems exhibited significant demographic disparities. Reports of these disparities have led to concerns among civil liberties groups and distrust of FRT in communities already subject to institutional bias and concerns about over-policing.
- *Privacy concerns about how face images are collected, used, and retained.* For example, some training data sets and reference databases (against which a candidate face image is matched) have been constructed from images scraped from social media and other online sources without any effort to gain consent from those pictured. Although such practices—for example, by Clearview AI—allowed large image databases to be amassed quickly, they have also raised privacy, fairness, and quality concerns. In addition, images captured in real time for recognition create risks of inappropriate data retention, secondary uses, and absent or insufficient opt-out procedures—and raise questions about whether and in what circumstances governments can acquire such information.

---

[11] K.S. Krishnapriya, K. Vangara, M. King, V. Albiero, and K. Bowyer, 2019, "Characterizing the Variability in Face Recognition Accuracy Relative to Race," pp. 2278–2285 in 2019 IEEE/CVF Conference on Computer Vision and Pattern Recognition Workshops (CVPRW), https://doi.ieeecomputersociety.org/10.1109/CVPRW.2019.00281.

[12] P. Grother, M. Ngan, and K. Hanaoka, 2019, "Face Recognition Vendor Test (FVRT): Part 3: Demographic Effects," National Institute of Standards and Technology, https://doi.org/10.6028/NIST.IR.8280.

[13] K. Krishnapriya, V. Albiero, K. Vangara, M.C. King, and K.W. Bowyer, 2020, "Issues Related to Face Recognition Accuracy Varying Based on Race and Skin Tone," *IEEE Transactions on Technology and Society* 1(1):8–20.

- *The use of substandard face recognition systems in high-stakes operational environments.* While a number of use cases for FRT may have low stakes for potential improper results, such as failing to unlock a personal device, other operational uses can result in serious consequences, such as false arrest and detention. The consequences of error thus vary tremendously by use case, making blanket rules or generalizations about use challenging.
- *Impacts on an array of privacy, civil liberties, and equity issues.* FRTs are highly personal, uniquely powerful, and potentially extremely intrusive. Opting out of showing one's face is not a realistic option in most circumstances. If used for real-time surveillance, FRTs can dramatically increase the scope and reduce the cost of collecting detailed information about a person's movements, activities, and associations. Without responsible guidelines, FRT systems permit easy open-ended collection of face data, even when the collecting institution has no particular person or incident as its focus. With respect to equity, just as unease about disproportionate surveillance in historically disadvantaged communities gives rise to concerns about the inequitable burden on those communities, disproportionate use of FRT would likewise raise concerns about an additional burden.
- *Potential risk of differential treatment due to the ease of identification.* The low cost of using FRT increases the ease with which persons may be identified for exclusion or rewards in ways that were once practicably impossible. For example, high-end retailers might use FRT to identify wealthy shoppers for preferential treatment or property owners might identify and deny entry to individuals who are not part of a protected class.
- *Compounding disadvantages.* For example, residents in public housing may be subject to enforcement of minor rules through the use of FRT originally intended to address public safety concerns.
- *Mass surveillance, political repression, and other human rights abuses.* If applied broadly and without safeguards, FRT allows repressive regimes to create detailed records of people's movements and activities, including political protests or organizing, and to block targeted individuals from participation in public life. This is not hypothetical; there is evidence of such use in multiple countries.

These concerns stem from an array of related characteristics of the technology, including the following:

- *Highly personal.* The face is a uniquely individualizing part of the body that is much more visible than other individualizing body parts as fingerprints or iris patterns. FRT can be operated at a significant distance and is inextricably tied to an individual in a way that other techniques, such as cell phone trackers or license plate readers, are not.
- *Pervasive.* FRT can exploit the large and growing number of images available from cameras operated by governments (e.g., cameras installed on city streets), businesses (e.g., security cameras), and individuals (e.g., doorbell cameras). Moreover, FRT can easily be applied after the fact to stored images and video. It is impossible or at least highly impractical to opt out of collection of face imagery by such devices—in contrast to alternatives such as cell phone tracking or license plate readers, where it is costly but not entirely impractical to opt out.
- *Ubiquitous.* Many if not most people can be recognized: in the United States most faces are all already in a government database and many people's labeled faces are available online. The technology is readily available to the private sector as well as to governments.
- *Stealthy.* It is difficult to detect whether FRT is being used in a given setting and for what purposes.
- *Inexpensive.* In contrast to human review of camera footage, FRT is automated. Relatedly, the marginal cost of using FRT is very low—unlike, for example, DNA testing, which still has a nontrivial per-use cost.

THE GOVERNANCE OF FACIAL RECOGNITION TECHNOLOGY

FRT raises novel and complex challenges for governance. The complexity arises because many legal and policy questions arise at points ranging from the development of the technology to deployment and use. There are distinct and unsettled legal and policy questions at numerous junctures, and governance of the technology will depend on both where and how FRT is used. Furthermore, the regulation of FRT might take place at different levels of government (i.e., national, state, and local), and at any given level, FRT might be subject to regulation by existing general laws (e.g., those related to intellectual property, privacy, law enforcement), technology specific law or regulation, or both.

Complicating this picture is the fact that, from a societal perspective, FRT is problematic because it impacts a core set of interests related to freedom from state and/or private surveillance, and hence control over personal information. Its use therefore has the ability to interfere with and substantially affect the values embodied in commitments to privacy, civil liberties, and human rights. Thus, there are multiple legislative and non-legislative approaches aimed at the governance of FRT. The following sections identify some recent efforts directed at the governance of FRT.[14]

Facial Recognition Technology Legislation in the United States

In the United States, no federal regulation currently imposes a general constraint on the public or private use of FRT. Several bills have been introduced in Congress to regulate FRT, but so far none have come up for a vote.

At the state level, several states have enacted broader privacy laws to protect biometric information. Illinois became, in 2008, the first state to enact legislation that regulates collection, use, safeguarding, and retention of biometric data.[15] Arkansas, California, Texas, and Washington subsequently enacted similar laws.

At the municipal level, the city of San Francisco became, in 2019, the first U.S. city to ban the use of FRT by its public agencies, including its police department, under its administrative code.[16] Other cities, including Oakland, California, and Somerville, Massachusetts, subsequently passed local ordinances restricting the use of FRT by public agencies. Since then, some have called for a reconsideration of these policies in light of concerns about crime.

Facial Recognition Technology Legislation Outside the United States

Perhaps most notably, the European Parliament recently passed the text of the Artificial Intelligence Act (the AI Act), an extensive and complex statute intended to regulate the development and use of artificial intelligence. The final text of the act had not been released as of this writing, but the act would complement the European Union's General Data Protection Regulation (GDPR)[17] and the Law Enforcement Directive of the European Union.[18]

Within the European Union, countries have also individually moved to regulate FRT. Countries around the world have also taken regulatory or other action on FRT (see Chapter 4).

---

[14] See Chapter 4 for additional details about such efforts.

[15] State of Illinois Biometric Information Privacy Act of 2008, Public Act 095-0994, 740 ILCS 14, effective October 3, 2008, https://www.ilga.gov/legislation/ilcs/ilcs3.asp?ActID=3004&ChapterID=57.

[16] See City and County of San Francisco, 2019, "Board of Supervisors Approval of Surveillance Technology Policy," Admin Code Section 19B.2(d), https://sfbos.org/sites/default/files/o0286-19.pdf.

[17] B. Wolford, ed., n.d., "What Is GDPR, the EU's New Data Protection Law," Proton Technologies AG, https://gdpr.eu/what-is-gdpr, accessed November 17, 2023.

[18] European Commission, "Data Protection in the EU," https://commission.europa.eu/law/law-topic/data-protection/data-protection-eu_en, accessed November 17, 2023.

Non-Legislative Governance Approaches

Many organizations have produced documents that recommend non-legislative governance approaches to the regulation of FRT. Such approaches often promote or identify norms, principles, or best practices that are encapsulated in, for example, codes of conduct, declarations, or guidelines. They may also take the form of directives.

In 2012, the U.S. Federal Trade Commission published a report titled *Facing Facts: Best Practices for Common Uses of Facial Recognition Technologies*.[19] The report detailed conversations that occurred during a 2011 workshop and coupled these with public commentary collected after the event.[20] The report identifies several best practices for FRT system design: (1) maintain reasonable data protection of consumers' face images and associated biometric data, (2) protect online face images from unauthorized collection, and (3) adopt appropriate retention and disposal practices for images of faces and other biometric data, and consider the sensitivity of information being collected or the sensitivity of the environment in which it is being collected. The report emphasizes transparency and affirmative consent as key factors in enabling consumers to make informed decisions about their data.

In 2017, the National Telecommunications and Information Administration released a report on the best privacy practices for commercial use of FRT.[21] Recognizing the growing use of the technology, the report emphasized the need for foundational guidelines to govern the use of FRT and application-specific best practices. It called for greater transparency about how and where FRT is being used and for policies to govern the collection, use, and storage of facial template data. In addition, the report offers principles to help organizations design policies to appropriately limit the use of face image data, implement adequate security safeguards, ensure image quality standards for their FRT systems, and develop appropriate procedures for problem resolution and redress.

As FRTs are widely used for criminal intelligence and investigations, in 2017, the National Criminal Intelligence Resource Center created, as part of a collaborative effort with multiple jurisdictions of law enforcement and civil liberties–focused actors, a template for FRT policy creation that focuses on privacy, civil rights, and civil liberties protection.[22] The template targets the collection, use, access, management, and destruction of FRT-related data and includes guidelines for accountability and enforcement.

---

[19] Federal Trade Commission, 2012, "Facing Facts: Best Practices for Common Uses of Facial Recognition Technologies," https://www.ftc.gov/sites/default/files/documents/reports/facing-facts-best-practices-common-uses-facial-recognition-technologies/121022facialtechrpt.pdf.

[20] Ibid.

[21] National Telecommunications and Information Administration, 2017, "Privacy Best Practices for Commercial Facial Recognition Use," https://www.ntia.doc.gov/files/ntia/publications/privacy_best_practices_recommendations_for_commercial_use_of_facial_recogntion.pdf.

[22] National Criminal Intelligence Resource Center, 2017, "Face Recognition Policy Template for State, Local, and Tribal Criminal Intelligence and Investigative Activities," https://bja.ojp.gov/sites/g/files/xyckuh186/files/Publications/Face-Recognition-Policy-Development-Template-508-compliant.pdf.

The template offers clear guidelines to help law enforcement in treating issues related to privacy, civil rights, and civil liberties through policy development and routine evaluation, training, review, and auditing. The effort aims to optimize FRT performance while emphasizing the need for careful and informed human oversight in cases that could particularly impact an individual's civil liberties.

A similar concern for safeguarding privacy, civil rights, and civil liberties motivated the White House Office of Science and Technology Policy to release a "Blueprint for an AI Bill of Rights in 2022."[23] The blueprint sets forth principles for the use of automated systems (including FRT): safe and effective systems; protection from algorithmic discrimination, data privacy, notice and explanation; and human alternatives, consideration, and fallback. For each principle, the blueprint sets forth baseline expectations for citizens and offers best practices to ensure that an automated system lives up to the vision of the AI Bill of Rights.

A report published by the Security Industry Association (SIA) builds on previous assemblages of FRT best practices,[24] emphasizing transparency, clearly stated purpose of use, human oversight, security, training for users and consumers, and privacy by design. The report offers guidelines for public-sector, private-sector, and law enforcement use of FRTs. It suggests that best practice requires that the best-performing FRT is used, proposing that, in addition to meeting standards set by organizations such as the National Institute of Standards and Technology, FRT users must mitigate against performance variability by employing the best current technology. The report emphasizes the importance of using FRT in a manner that does not discriminate, suggesting that the current highest-performing FRTs have been shown to have equal performance across demographic groups.

The Department of Homeland Security (DHS) recently issued Directive Number 026-11, titled *Use of Face Recognition and Face Capture Technologies*. The directive reiterates that FRT is "only authorized for use for DHS missions, in accordance with DHS' lawful authorities" and that it is critical that DHS only use FRT "in a manner that includes safeguards for privacy, civil rights, and civil liberties." It requires, among other things, that FRT be independently tested and evaluated; that, when FRT is used for verification for non-law-enforcement-related actions or investigations, an opt-out and alternative processing is available; that alternative processing is available to resolve match or no match outcomes; and that FRT "used for identification may not be used as the sole basis for law or civil enforcement related actions, especially when used as investigative leads."

---

[23] Office of Science and Technology Policy, n.d., "Blueprint for an AI Bill of Rights: Making Automated Systems Work for the American People," https://www.whitehouse.gov/ostp/ai-bill-of-rights, accessed May 23, 2023.

[24] Security Industry Association, 2020, "SIA Principles for the Responsible and Effective Use of Facial Recognition Technology," https://www.securityindustry.org/report/sia-principles-for-the-responsible-and-effective-use-of-facial-recognition-technology.

Furthermore, "any potential matches or results from the use of" FRT are to be "manually reviewed by human face examiners prior to any law or civil enforcement action."[25]

ABOUT THIS REPORT

Mindful of the potential uses of FRTs and the associated potential concerns outlined in this chapter, DHS's Office of Biometric Identity Management and the Federal Bureau of Investigation commissioned this report to assess current capabilities, future possibilities, societal implications, and governance of FRTs. The study committee appointed by the National Academies of Sciences, Engineering, and Medicine, which wrote this report, was tasked with reviewing current use cases; explaining how FRTs operate; and considering the legal, social, and ethical issues implicated by their use. See Appendix A for the full statement of task.

The remainder of this report addresses these issues.

Chapter 2 looks at the current state of FRT, placing today's state-of-the-art systems in historical context and explaining the relationship of FRT to other emerging technologies such as AI and providing an overview of performance measurement and trends.

Chapter 3 provides an overview of major use cases of the technology. It uses brief vignettes to illustrate both current and potential use cases and some of the potential benefits and concerns they present.

Chapter 4 reviews concerns raised by the use of FRT, particularly with regard to equity, privacy, and civil rights, and examines how these concerns affect the governance of FRT.

Chapter 5 describes policy options and presents the committee's conclusions and recommendations along with an initial sketch of a risk management framework designed to help organizations think through best practices for different types of use cases.

---

[25] Department of Homeland Security, 2023, "Use of Facial Recognition and Face Capture Technologies," https://www.dhs.gov/sites/default/files/2023-09/23_0913_mgmt_026-11-use-face-recognition-face-capture-technologies.pdf.

2
Facial Recognition Technology

Development of facial recognition technology (FRT) began around 1970.[1] In the past decade, the pace of development has accelerated with the industrial adoption and adaptation of various neural network–based machine learning techniques. These advances have led to remarkable gains in recognition accuracy and speed.

Specifically, when photographs are acquired cooperatively and under constrained conditions—such as in passport or driver's license applications or when crossing an international border—the photos are of sufficient quality to support high-confidence, high-accuracy retrieval from databases of such photographs. Using the leading 2023 face recognition algorithms to search a mugshot database of 12 million identities, fully 99.9 percent of searches will return the correct matching entry.[2] The only failures result from changes in facial appearance associated with acute facial injury and long-term aging. This one result, however, involved the use of photos taken under mostly ideal conditions in which the photography is formally standardized, and the subject cooperates with the photographer. If those conditions do not apply, accuracy falls off sharply. Between these two extremes, accuracy will vary and any measurement of it must be accompanied by a narrative about how the photos were acquired.

The potential for very high accuracy must be further qualified by considerations of what the FRT is used for, and on whom:

---

[1] T. Kanade, 1973, "Picture Processing System by Computer Complex and Recognition of Human Faces," https://repository.kulib.kyoto-u.ac.jp/dspace/bitstream/2433/162079/2/D_Kanade_Takeo.pdf.

[2] P.J. Grother, M. Ngan, and K. Hanaoka, 2019, *Face Recognition Vendor Test (FRVT)—Part 2: Identification*, Washington, DC: Department of Commerce (DOC) and Gaithersburg, MD: National Institute of Standards and Technology (NIST), https://www.nist.gov/system/files/documents/2019/09/11/nistir_8271_20190911.pdf.

- Many applications require correct rejection of faces that are not in the database—that is, the minimization of false matches. This is critical to avoid identity mismatches that can, in certain applications, have adverse consequences for an individual's civil liberties.
- Error rates are not always the same for all queries; they can vary by demographic group or even person to person, but these variations are becoming smaller as FRT models continue to evolve.
- Identical twins represent an extreme yet realistic example of persons who may cause false matches. Approximately 0.4 percent of births in the United States are identical twins.[3]
- Some use cases require a search to produce high-confidence matches—where the face recognition software deems the match to be highly similar—that is, above some minimum pairwise similarity threshold.
- Some applications include mechanisms to detect subjects trying to impersonate someone else or to conceal their own identity—for example, by wearing makeup or wearing a high-quality silicone mask. These evasion-detection mechanisms do not always work and can contribute to errors.

The frequency of errors always depends on the design and engineering of the system. The consequences of errors depend on how the system is used. This chapter begins by describing the algorithms, image capture hardware, and performance improvements over time. It then turns to pose, illumination, expression and facial aging challenges, demographic effects, and sources and consequences of errors. It concludes by looking at human examiner roles and capabilities and several salient attributes of commercially deployed FRT systems.

ALGORITHMS

A face recognition algorithm has three parts: a detector, a feature extractor, and a comparator. The detector will find a face in an image; perhaps rotate, center, and resize it; and produce an image suitable for feature extraction. The feature extraction step, known more generically as *template generation*, performs various elaborate computations on the pixel values, and produces a set of numbers that are known in various communities

---

[3] P. Gill, M.N. Lende, and J.W. Van Hook, 2023, "Twin Births," updated February 6, In *StatPearls [Internet]*, Treasure Island, FL: StatPearls Publishing, https://www.ncbi.nlm.nih.gov/books/NBK493200.

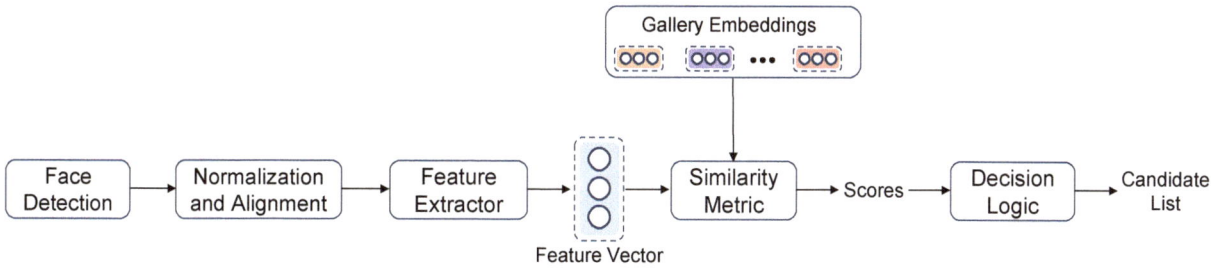

FIGURE 2-1 Face recognition pipeline.
NOTE: Many systems will only output a ranked list of candidates with similarity scores above a specified threshold.

as a template, a feature vector, or an embedding; this report uses the term "template."[4] This process is depicted in Figure 2-1.

A template is designed to support the core recognition goal of the comparator, which takes two templates and produces a single number expressing how similar the faces were that went into the templates. Comparison code is usually quite simple. The result is universally known as a *similarity score*, often normalized between 0 and 1. If the score is high, this is taken as an indication that the two input faces were from the same person. This interpretation is discussed further in the following text.

Face recognition can thus be used to confirm identity to authenticate that a user is who they claim to be. Similarity scores are not the same as personal identification numbers (PINs) and passwords, which authenticate a user only if they are identical to what the user initially specified. Note that similarity scores are used rather than a binary match/no match because no two photos of a face are identical—owing to even the slightest variations in lighting, facial expression, head position, and camera noise.

The task of FRT is to ignore the "nuisance" variations in face images such as those shown in Figure 2-2 and produce templates that when compared yield high similarity scores from the photos of the same person and low scores for photos of different people. This task is the core difficulty in improving FRTs' accuracy; it is addressed today by training a neural network to learn from many highly variable images in terms of pose, illumination, expressions, aging, and occlusion of many people, ranging from tens of thousands to tens of millions. Such images are usually of real individuals, but in recent years there is considerable interest in synthesizing face images in unlimited quantities using a different class of neural network[5] to increase the size of the training data.

---

[4] The term "faceprint" has been used, but this should be deprecated because its progenitor "fingerprint" applies to an image, not to features derived from it.
[5] P. Melzi, C. Rathgeb, R. Tolosana, et al., 2023, "GANDiffFace: Controllable Generation of Synthetic Datasets for Face Recognition with Realistic Variations," arXiv:abs/2305.19962.

 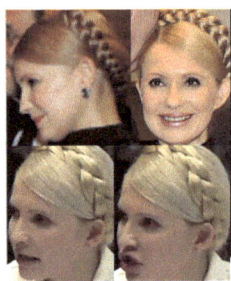  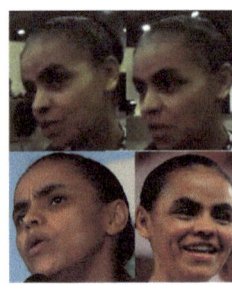 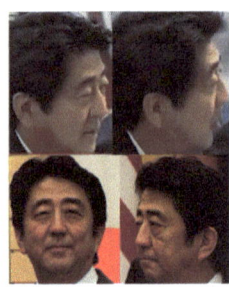

FIGURE 2-2 Examples of variations in the face images of the same person that could alter a similarity score.
NOTE: These variations include full pose variation, a mixture of still images and video frames, and a wide variation in imaging conditions and geographic origin of subjects.
SOURCE: © 2015 IEEE. Reprinted, with permission, from B.F. Klare, 2015, "Pushing the Frontiers of Unconstrained Face Detection and Recognition: IARPA Janus Benchmark A," *2015 Conference Proceedings on Computer Vision and Pattern Recognition (CVPR)* 1931–1939.

History

Although human beings have been using faces to recognize one another since time immemorial,[6] the work on enabling computers to recognize human faces was started in the mid-1960s by Woodrow W. Bledsoe and his colleagues at Panoramic Research. Bledsoe qualified his face recognition system as a "man-machine" system, because it required human experts to first manually locate some facial landmarks on a photograph. The comparison was then performed automatically based on 20 normalized distances derived from these facial landmarks (e.g., width of the mouth, width of eyes, etc.). Bledsoe observed that "[t]his recognition problem is made difficult by the great variability in head rotation and tilt, lighting intensity and angle, facial expression, aging, etc."[7]

A method to automatically extract such facial landmarks was first proposed in Takeo Kanade's 1973 PhD thesis, which can be considered to have presented the first fully automatic FRT system.[8] Although the earliest face recognition systems were based on geometric features (distances between pre-defined landmarks), Sirovich and Kirby in 1987[9] and later Turk and Pentland in 1991 showed that faces could be represented by extracting features from all the pixels in the whole image by a method known as principal component analysis.[10] This holistic appearance-based technique generates a

---

[6] Adapted in part from A.K. Jain, K. Nandakumar, and A. Ross, 2016, "50 Years of Biometric Research: Accomplishments, Challenges, and Opportunities," *Pattern Recognition Letters* 79(3.2):80–105, https://doi.org/10.1016/j.patrec.2015.12.013.

[7] W.W. Bledsoe, 1966, *Man-Machine Facial Recognition: Report on a Large-Scale Experiment*, Palo Alto, CA: Panoramic Research, Inc.

[8] T. Kanade, 1974, *Picture Processing System by Computer Complex and Recognition of Human Faces*, https://repository.kulib.kyoto-u.ac.jp/dspace/bitstream/2433/162079/2/D_Kanade_Takeo.pdf.

[9] L. Sirovich and M. Kirby, 1987, "Low-Dimensional Procedure for the Characterization of Human Faces," *Journal of the Optical Society of America A* 4(3):519, https://doi.org/10.1364/josaa.4.000519.

[10] M. Turk and A. Pentland, 1991, "Eigenfaces for Recognition," *Journal of Cognitive Neuroscience* 3(1):71–86, https://doi.org/10.1162/jocn.1991.3.1.71.

compact representation of the entire face region in the acquired image. As an example, a 64 × 64 pixel face image (a total of 4,096 pixels) could be represented in terms of merely 100 feature values that are learned using a training set of face images. These features have the property that they could be used to reconstruct the original face image with sufficient fidelity. Two other historical examples of face recognition approaches are the local feature analysis method of Penev and Atick and the Fisherface method of Belhumeur et al.[11,12]

Model-based techniques derive a pose-independent representation by building two-dimensional or three-dimensional models of the face. They generally rely on detection of several fiducial points in the face such as the chin, the tip of the nose, the corners of eyes, or the corners of the mouth. The pioneering work in this area was Wiskott et al.'s elastic bunch graph matching approach.[13] Another advance, which uses three-dimensional models and both facial texture and shape features, is the morphable model proposed by Blanz and Vetter.[14]

Appearance-based schemes use raw pixel intensity values and are thus very sensitive to variations in ambient lighting and facial expression. Texture-based methods such as scale-invariant feature transform[15] and local binary patterns[16] were developed to reduce that sensitivity. These methods make use of more robust representations that characterize image texture using the distribution of local pixel values rather than individual pixel values.

Most face recognition techniques assume that faces can be aligned and properly normalized geometrically and photometrically. Alignment is typically performed using the location of the two eyes in a face. The face detection scheme developed by Viola and Jones[17] was a milestone because it enables faces to be detected in real time even in the presence of background clutter, a situation commonly encountered in surveillance applications. Even though the Viola–Jones detector performs very well in real-time applications, it struggles with illumination changes, non-frontal facial poses, and occlusion—and is thus outdated.

---

[11] P.S. Penev and J.J. Atick, 1996, "Local Feature Analysis: A General Statistical Theory for Object Representation," *Network: Computation in Neural Systems* 7(3):477–500.

[12] P.N. Belhumeur, J.P. Hespanha, and D.J. Kriegman, 1997, "Eigenfaces vs. Fisherfaces: Recognition Using Class Specific Linear Projection," *IEEE Transactions on Pattern Analysis and Machine Intelligence* 19(7):711–720.

[13] L. Wiskott, J.-M. Fellous, N. Krüger, and C. Von Der Malsburg, 2022, "Face Recognition by Elastic Bunch Graph Matching," Pp. 355–396 in *Intelligent Biometric Techniques in Fingerprint and Face Recognition*, New York: Routledge.

[14] V. Blanz and T. Vetter, 2003, "Face Recognition Based on Fitting a 3D Morphable Model," *IEEE Transactions on Pattern Analysis and Machine Intelligence* 25(9):1063–1074.

[15] D.G. Lowe, 1999, "Object Recognition from Local Scale-Invariant Features," *Proceedings of the International Conference on Computer Vision* 2:1150–1157.

[16] T. Ojala, M. Pietikainen, and D. Harwood, 1994, "Performance Evaluation of Texture Measures with Classification Based on Kullback Discrimination of Distributions," *Proceedings of 12th International Conference on Pattern Recognition* 1:582–585.

[17] P. Viola and M.J. Jones, 2004, "Robust Real-Time Face Detection," *International Journal of Computer Vision* 57(2):137–154.

Artificial Intelligence–Based Revolution

Over the past decade, the field of face recognition has significantly advanced, primarily due to breakthroughs in an artificial intelligence technique known as deep convolutional neural networks (DCNNs), which were originally developed for optical character recognition and later applied to diverse computer vision tasks such as automated driving and medical image analysis. These deep learning techniques have proven to provide the most prominent advance in face recognition.

The application of DCNNs to face recognition was demonstrated to great effect in 2014, when researchers at Facebook trained a network with between 800 and 1,200 photos of each of 4,030 persons to obtain greatly improved accuracy on the open benchmark data sets of the day.[18] The performance gains stemmed from increased tolerance of nuisance properties of image invariance to facial appearance variations that are extraneous to the identity of the subject. It remained to be seen whether that class of algorithm could also learn to distinguish between individuals in much larger populations than the 4,030 that Facebook used, a requirement because even before 2014, face recognition algorithms were being applied to populations of tens of millions. Ultimately, Facebook's approach—leveraging larger numbers of photos from social media—proved revolutionary for the wider biometrics industry: over the next decade, the suppliers of face recognition algorithms largely discarded their prior hand-crafted feature techniques and adopted the new DCNN methods, adapting, modifying, and expanding them as an enormous research community developed the new technologies. Research since 2014 has further evolved the DCNN-based approach.[19] A 2019 paper described significant improvements to the design of loss functions for face recognition.[20] A well-maintained Git repository[21] contributes to the popularity of this work in the computer vision community and has helped make it the "go to" approach in face recognition and establish it as a new baseline. It has received more than 5,600 citations since its publication.

The deep neural network approach is illustrated in Figure 2-3. Once a face has been detected in its parent image, it will usually be rotated, cropped from its parent image, and then resized to the size of the input layer of the neural network. Some

---

[18] Y. Taigman, M. Yang, M. Ranzato, and L. Wolf, 2014, "Deepface: Closing the Gap to Human-Level Performance in Face Verification," *2014 IEEE Conference on Computer Vision and Pattern Recognition,* pp. 1701–1708.

[19] Some of this work was supported by the Intelligence Advanced Research Projects Agency through the JANUS program that ran from 2014 to 2020. See Office of the Director of National Intelligence, "JANUS," https://www.iarpa.gov/research-programs/janus, accessed November 17, 2023.

[20] J. Deng, J. Guo, N. Xue, and S. Zafeiriou, 2019, "ArcFace: Additive Angular Margin Loss for Deep Face Recognition," *Proceedings of the IEEE/CVF Conference on Computer Vision and Pattern Recognition (CVPR)* 4690–4699, https://doi.org/10.1109/CVPR.2019.00482.

[21] J. Guo and J. Deng, 2021, "ArcFace with Parallel Acceleration on Both Features and Centers, Original MXNet Implementation on InsightFace," GitHub, https://github.com/deepinsight/insightface/tree/master/recognition/arcface_mxnet.

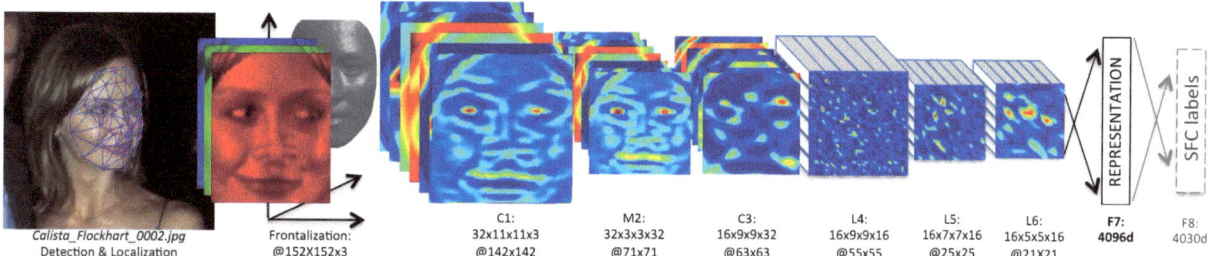

FIGURE 2-3 Images that contain coarse patterns extracted from the input.
SOURCE: © 2014 IEEE. Reprinted, with permission, from T. Yaniv, M. Yang, M. Ranzato, and L. Wolf, 2014, "DeepFace: Closing the Gap to Human-Level Performance in Face Verification," *2014 IEEE Conference on Computer Vision and Pattern Recognition (CVPR)* 1701–1708.

developers may perform these steps in a different order. Some developers may apply various image processing steps also—for example, to brighten the image. The input layer of the neural network is quite small—say, 112 × 112 pixels or 256 × 256 pixels—and usually square. This has implications, as discussed later.

The DCNN accepts the input image, usually as a color image with red, green, and blue color channels, and feeds it forward through a many-layered computation. In the first layer, the pixels are weighted and averaged and combined in many ways, the net effect of which is to produce a set of somewhat smaller-size outputs that can be viewed as images that contain coarse patterns extracted from the input (see Figure 2-3).

This output is then passed through a non-linear function, a necessary hallmark of neural computation. The second layer proceeds with a slightly different set of weights and computations, and its output is again transformed non-linearly. The layered computation continues with each output, when visualized, being a more abstract, less human-interpretable, version of the input face image. The feed-forward process culminates with the production of a vector, a set of numbers that comprises the template. The set of numbers is included in the biometric template, perhaps along with bookkeeping information such as the date, and the version of the DCNN.

Templates are generally reversible—they do not provide the privacy benefits afforded by one-way hashes; they can be reversed, with some difficulty, to something with some resemblance to the original face.[22] They can also leak other information about an individual such as sex. As a result, templates must also be protected from disclosure in order to protect individual privacy.

---

[22] See A. Zhmoginov and M. Sandler, 2016, "Inverting Face Embeddings with Convolutional Neural Networks," arXiv preprint, arXiv:1606.04189; or G. Mai, K. Cao, P.C. Yuen, and A.K. Jain, 2019, "On the Reconstruction of Face Images from Deep Face Templates," *IEEE Transactions on Pattern Analysis and Machine Intelligence* 41(5):1188–1202.

There is considerable variation in template generation speed across today's algorithms, with accurate algorithms producing templates from 0.1 second to several seconds on a server-class CPU. Faster algorithms can be ported to run on processors embedded in cameras or physical access-control devices. Graphical processing units (GPUs) that are considered essential for training algorithms are typically not necessary for recognition. When FRT is applied in video feeds, or when many images are captured, or when many faces appear in an image, a GPU may be employed to provide real-time recognition.

Resolution

Contemporary face recognition algorithms operate at very low resolution. They typically operate on face photographs that have been cropped and resized so that the head and face fill an image of size 112 × 112, 128 × 128, or 256 × 256 pixels. These sizes mean that the inputs to the algorithms have resolution low enough that it will not be possible to see human hair, skin pores, and similar-size detail. Operators of face recognition often cite standards that mandate collection of larger images, but core algorithms operate at a size determined by developers. Sizes are much smaller than the images collected by contemporary mobile phones or digital cameras (e.g., 3,000 × 4,000 pixels). They are also much smaller than the images preferred by the community of forensic examiners who review face pairs and testify in court. Human reviewers find value in high-resolution images because they support exculpation: if a specific feature is visible in one photo but not the other, this can be dispositive.

For example, Figure 2-4 shows how scars and moles could enable a reviewer to correctly distinguish between identical twins. Such marks are often not present in younger twins. Also, such fine details are typically not used by automated algorithms because they are often not visible in low-resolution images. These issues argue for the wholesale migration of the industry to high-resolution images, something that is not readily achieved because such images are not available to developers of FRT algorithms in sufficient quantities for training DCNNs.

Template Extraction Model Training

The models in face recognition algorithms convert an image to a template. The models are usually trained in the developer's research and development laboratories; each developer uses different variants of DCNN and training protocols and has access to different training sets. Furthermore, these models are rarely trained on data derived from the operational environment where the system is ultimately deployed. Therefore, the characteristics of the images in the data set used to train the FRT model may differ from those encountered in an operational setting.

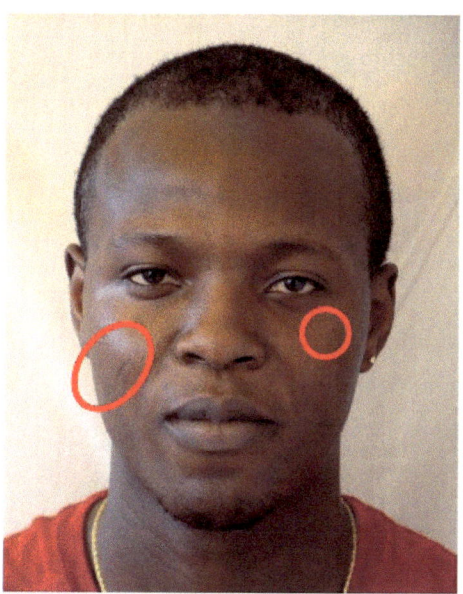 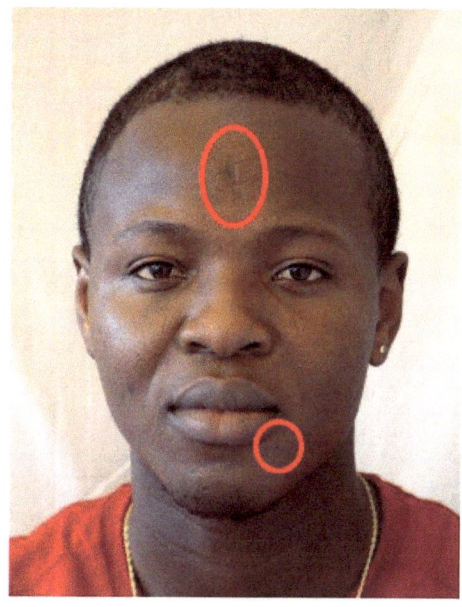

FIGURE 2-4 Highlighted unique identifiers in two portrait photographs of identical twins.
SOURCE: © 2011 IEEE. Reprinted, with permission, from B. Klare, A.A. Paulino, and A.K. Jain, 2011, "Analysis of Facial Features in Identical Twins," *2011 International Joint Conference on Biometrics (IJCB)* 1–8.

Training is key to the performance of the algorithm, and much of the intellectual property resides in the expert curation of data sets, selection of architecture, specification of loss functions, intervention, and selection and tuning of parameters. The training is almost always supervised, a term borrowed from machine learning that means that each training sample (face image) has an identity label associated with it. Thus, during training the DCNN learns to associate face images of the same identity and simultaneously to distinguish between faces of different identities and does this with low classification error. It is of commercial value therefore for a developer to possess, or have access to, a large number of face images and their associated identity labels. Such databases should come from a large number (millions) of individuals and have a large number (thousands) of diverse images per individual. The identity labels must have high integrity—the person in the image must be correctly labeled. It is costly to procure such a large collection of labeled photos, an expense that was historically avoided by many researchers by collecting photos from the Web—the popular Labeled Faces in the Wild[23] and MS-Celeb[24] databases were assembled in this way—and several such databases have

---

[23] G.B. Huang, M. Ramesh, T. Berg, and E. Learned-Miller, 2007, *Labeled Faces in the Wild: A Database for Studying Face Recognition in Unconstrained Environments*, Technical Report 07-49, Amherst: University of Massachusetts.

[24] Y. Guo, L. Zhang, Y. Hu, X. He, and J. Gao, 2016, "MS-Celeb-1M: A Dataset and Benchmark for Large-Scale Face Recognition." Pp. 87–102 in *Computer Vision–ECCV 2016*, European Conference on Computer Vision, Lectures Notes in Computer Science, Vol. 9907, Cham, Switzerland: Springer.

since been expunged due to privacy and ethics concerns. Ironically, the Diversity in Faces database,[25] which was assembled to support development of equitable face analysis algorithms, was withdrawn as the collection of images from the Web quickly became controversial. The database was not suitable for development of actual recognition algorithms because it did not include identity labels.

Comparison and Similarity Scores

The final step in the face recognition algorithm is to compare two templates. The comparison module is often a simple piece of code that accepts two templates and computes some measure of how similar they are. This is known as one-to-one comparison. The method is generally a trade secret, but it generally treats the templates as vectors in a notional high-dimensional space and measures distance as a Euclidean distance (as the crow flies), a Manhattan distance (walking city blocks), or simply the angle between these vectors. If the distance measure is small (or equivalently the similarity measure is high), then it is likely that the two photos are of the same face (see also the discussion of errors in the section "Accuracy"). By industry convention, such numbers are presented on a similarity scale, where bigger values connote similarity of the faces.

Although high similarity scores are often construed to indicate sameness of identity of faces in two photographs, a low score should not be taken to be a definitive statement that two faces are from different people. The key factor is photo quality. Consider a comparison of two photos of the same person—a passport-style photo compared with an image of a face captured from a camera whose lens was far from well focused. The second photo has low resolution or information content such that most face recognition algorithms will return a low similarity score, just as they would from comparison of two high-information content passport photos of unrelated people. Thus, low scores stem from either a difference in identity or low image quality.

Importantly, *similarity scores* cannot be interpreted as likelihoods, probabilities, or a "percentage match." This is true because each developer emits scores on their own proprietary interval; it is common to use [0,1], [0,100], but others use [0,19000], [2,3], and [0.6,0.9]. The distribution of scores within those intervals will vary by developer: some give continuous normal-like distributions; others arrange to pin non-mate and mate scores to 0 and 1, or 0 and 100, respectively. As such, there is no universal interpretation of when a similarity between two faces is "strong"—that is, high enough to confirm that two photos are of the same person. Nevertheless, such interpretations are sometimes made by system operators, and this can prejudice or bias human review of

---

[25] M. Merler, N. Ratha, R.S. Feris, and J.R. Smith, 2019, "Diversity in Faces," arXiv:1901.10436.

images.[26] There are no standards governing score values or statistical properties of similarity scores.

One-to-Many Identification

The larger and more demanding uses of face recognition involve search, known as one-to-many identification. Such applications first construct a template from "probe" imagery and then search it in a collection of previously enrolled templates known as a reference database or gallery. This operation is useful because the gallery entries are accompanied by some metadata—for example, a name, a location, or a URL—so that a successful search can yield some knowledge about the person in the search photo. It is very commonly implemented by comparing the probe's template with each enrollment template, followed by a sort operation that ranks and returns the most similar enrollments.

There are also algorithms that use alternative approaches to a series of one-to-one comparisons with each template in a gallery. Some use *fast search* algorithms, which afford extremely rapid search but with one-time expense of building a data structure such as a tree, graph,[27] or a dictionary. Others use a prebuilt data structure to provide better demographic stability. These algorithms, which represent a sizable minority of all search algorithms, do not yield the same scores as performing the series of one-to-one comparisons.

Some search algorithms are built to give sublinear search time. This means that if the number of images enrolled into a reference database is increased 100-fold, the search duration may only grow by, say, 2-fold. Such systems are characterized by very fast search. One highly accurate algorithm submitted to the National Institute of Standards and Technology's (NIST's) Facial Recognition Vendor Test performs a search of a 12-million-entry database in a few tens of milliseconds on a commodity CPU. Such capability, without any loss in search accuracy, is essential to practical applications in which many faces are searched against potentially large databases. The alternative, to use a linear search algorithm, would require more hardware resources.

---

[26] J.J. Howard, L.R. Rabbitt, and Y.B. Sirotin, 2020, "Human-Algorithm Teaming in Face Recognition: How Algorithm Outcomes Cognitively Bias Human Decision-Making," *PLOS ONE* 15(8), https://doi.org/10.1371/journal.pone.0237855.

[27] Y.A. Malkov and D.A. Yashunin, 2020, "Efficient and Robust Approximate Nearest Neighbor Search Using Hierarchical Navigable Small World Graphs," *IEEE Transactions on Pattern Analysis and Machine Intelligence* 42(4):824–836.

IMAGE ACQUISITION

Practical face recognition systems have also benefited from improvements in camera resolution and resulting image quality.

Cameras

The role of the camera as part of a face recognition system is to provide an image suited to the recognition process. The appearance of such images has been formally standardized since the 1990s, and de facto standardized since faces were collected in the criminal justice system more than a century ago and printed on international travel documents started after World War II. Today, the standard face appearance is specified by the ISO/IEC 39794-5:2019 standard,[28] which defines a placement geometry and frontal viewpoint as illustrated in Figure 2-5, and requires the absence of blur, shadows, occlusion, and areas of under- or overexposure.

The availability of low-cost, compact, and high-resolution cameras that can be embedded in various devices has been a key enabler of real-time and accurate FRT systems.

A key turning point in camera technology was the commercialization of digital cameras in the early 1990s. The frame rate, pixel density, and pixel sensitivity of image sensors have improved significantly. At the same time, image sensors have become smaller and cheaper, and good-quality face images can be captured today using smartphones or wearable devices. Low-cost cameras, such as Microsoft's Kinect, that can capture three-dimensional images in real time also entered the commercial market.

Cameras in use today range from inexpensive webcams to long-range surveillance cameras—and despite the overall improvements described here produce images that cover a wide range of quality. They are differentiated by several technical factors. First is whether they furnish a single image (stills) or a video stream. Stills are used in many applications, such as capturing a passport photo, while videos are naturally produced in settings where continuous imaging is in use, such as a closed-circuit television (CCTV) security camera. A second technical factor is whether the camera has any built-in capability for detecting faces.

Almost all security cameras, body-worn cameras, and ATM cameras observe and record scenes without specifically detecting and recognizing faces, which typically undermines image quality and face recognition accuracy. On the other hand, mobile phones are often equipped with cameras that will detect a face in a scene and, assuming

---

[28] International Organization for Standards (ISO), 2019, "Information Technology—Extensible Biometric Data Interchange Formats—Part 5: Face Image Data," ISO/IEC 39794-5:2019.

FIGURE 2-5 Example of the standard face appearance.
SOURCE: P. Grother, M. Ngan, and K. Hanaoka, 2019, *Face Recognition Vendor Test (FRVT), Part 3: Demographic Effects*, NISTIR 8280, Washington, DC: National Institute of Standards and Technology, Department of Commerce, https://nvlpubs.nist.gov/nistpubs/ir/2019/NIST.IR.8280.pdf.

that is the object of interest, focus and correctly expose that face. Such face-aware capture, although intended for aesthetic reasons, will improve face recognition accuracy essentially as a by-product. Mobile phone camera quality benefits also from high-dynamic-range sensors and the use of computational photography techniques.

Specifying the correct camera for an application is usually not sufficient to ensure accuracy because the environment in which it is used influences the properties of images. For example, if a camera is placed facing a window, subjects' faces can be underexposed. Similarly, if a building access control system is equipped with a camera expected to operate at night, then supplemental illumination will be necessary. There are many applications that allow for the deployment of face recognition systems in environments that support high accuracy.

Image Quality Assessment

Some systems incorporate quality assessment (QA) software that analyzes a photograph and quantifies whether it is in some sense acceptable. There are several use cases for such a capability—all are intended to improve the quality and thereby the likelihood that downstream recognition will succeed. A primary role for QA software is to detect a poor photograph and immediately prompt the subject or the photographer to take a better photograph. The software sometimes offers specific feedback on how to correct the problem. Typical problems include blur owing to motion; the subject not facing the camera; part of the face not visible owing to the subject wearing a cap, scarf, sunglasses, or the like; or the subject presenting a non-neutral expression.

Presentation Attack Detectors

In applications of face recognition that confer some benefit to the subject, there may be an incentive for a bad actor to attempt impersonation—that is, to fool the system into affirming a match to a falsely claimed identity. This deception is commonly attempted by presenting a printed photo or tablet display, or by wearing a face mask. Presentation attack detectors (PADs) are intended to thwart such attacks. They consist of software and sometimes hardware intended to generate additional signals for analysis.

In other applications, where a subject is motivated to not be recognized by a system, they may alter their appearance—for example, by wearing a disguise or a mask, or by presenting a photo of someone else. Again, the PAD system is intended to detect the subversive attempt.

POSE, ILLUMINATION, EXPRESSION, AND FACIAL AGING EFFECTS

Technological advancements have progressively tackled challenges caused by variations associated with pose, illumination, and expression. Contemporary algorithms are trained to tolerate such appearance changes, and also to handle changes inherent in facial aging. This is achieved, as mentioned earlier, by DCNNs that extract from photographs only the information that is salient to identity and ignore these so-called nuisance variations.

To demonstrate insensitivity to such extraneous factors, consider the following photo search results. When the image shown in Figure 2-6(A) is placed in a database with mugshots of 12 million other adult individuals, many recent face recognition algorithms correctly return it as the most similar face when searched with any of the photos shown in Figure 2-6(B). Those photos, taken from 2 to 19 years later, exhibit various changes in facial appearance—see the captions—that until the current decade would have mostly proved fatal to recognition retrieval.

The search accuracy described earlier has enabled many commercial and law enforcement applications—for example, detection of duplicate driver's license photos. Although the population of six of the U.S. states exceeds the 12 million used here, the technology remains viable in much larger populations, with a retrieval rate or search accuracy that declines slowly with increase in gallery size, as discussed later. This success, analogous to the needle-in-the-haystack problem, has limits. If the quality of the probe photograph is sufficiently degraded, as in Figure 2-6(C), the search will fail. For that image, all but one algorithm used in an NIST test fails to find the true match.

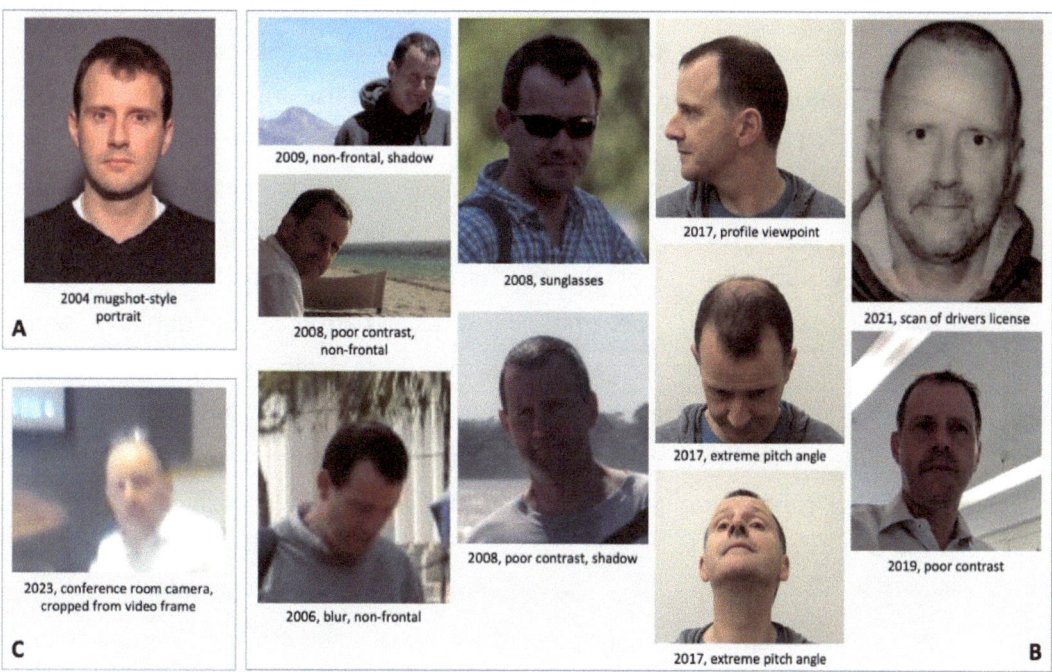

FIGURE 2-6 (A) Original image; (B) examples of changes in facial appearance that modern algorithms can correctly match; and (C) an example of a degraded image for which a search will fail.
SOURCE: P. Grother, with permission.

Two algorithms in the NIST test yielded partial success: One found the match, but judged 15 of the 12 million non-matching photographs to be more similar—that is, returned a rank 16 match. The 15 more-similar candidate identities are false matches—instances where the wrong identity is returned. A second algorithm gave the match at rank 42. These two outcomes show the power of the technology near its limit. The two algorithms can discern enough information from a heavily blurred photo to allow top 50 retrieval in a database of size 12 million.[29]

These two outcomes show why law enforcement investigators find extraordinary value in FRT; they potentially get a lead that, without FRT, they would not have. The fact that both algorithms yielding a match did so with a high rank is problematic in that a human reviewer must exonerate the other candidate identities. This point is discussed further in the section "Demographic Disparities."

High-rank hits (i.e., low similarity values for true matches) were much more common a decade ago even with better-quality search photos because the algorithms then could not discern information in a photograph to support assignment of high scores

---

[29] National Institute of Standards and Technology, 2020, "Face Recognition Vendor Test (FRVT)," updated November 30, https://www.nist.gov/programs-projects/face-recognition-vendor-test-frvt.

to true matches. Today, a high proportion of searches will return the correct match at rank 1. However, when the query face quality is low, and face aging has occurred (i.e., a large time lapse between the search photo and its true mate in the database), the true matches will have low similarity, comparable to those of false matches. These outcomes can present operational problems, because there is no clear result for the search photo. The impact of such outcomes depends on how the technology is used.

The primary source of false matches is when the person in a search photo has no match in the database. For example, most casino patrons would not be present in the establishment's compulsive gamblers or card-sharp databases. To suppress false positives (FPs) in such applications, a face recognition system for this application should be configured to return only highly similar candidates. If one is returned, further action is implied, either taking another photo and searching again, or involving a human to review the candidate identity.

This is a difficult task, as discussed later, made more difficult because of facial similarity that occurs naturally, particularly in twins and other siblings. As an example, a photograph of one adult sister was placed along with 12 million unrelated photos, and the resulting database searched with a photo of the other sister. All algorithms tested returned the sister as the most similar match. The similarity score was lower but still higher than those from searches of unrelated individuals. The system could be configured to correctly reject the sister in this instance, but that would not be effective for identical twins, who almost always produce high-scoring false matches.[30]

The approach of configuring a similarity threshold is unusual in criminal investigations. There, a face recognition search always returns a list of the most similar candidate photos. These are presented to police officials, in order of similarity to the search photo, for review in a bid to determine the identity of a face in an unknown photograph. The system is configured without a threshold, so the algorithm returns candidates whether the subject is in the database or not. By employing a human reviewer to compare photos and make decisions, accuracy becomes dependent on both the algorithm and the human. This has important consequences, as discussed here.

To see why face recognition is used in this way, consider the investigation of the Boston Marathon bombing.[31] There, authorities attempted to determine identities of all onlookers. Face recognition was used, and while it did not prove fruitful at that time, the motivation was clear. If one were to repeat two of the searches with present-day algorithms, the investigation might have been different. Repeating the demonstration here, when the Figure 2-7(A) photo of the convicted bomber is placed into a

---

[30] Ibid.

[31] J.C. Klontz and A.K. Jain, 2013, "A Case Study on Unconstrained Facial Recognition Using the Boston Marathon Bombings Suspects," Technical Report, Michigan State University.

FIGURE 2-7 Database image and example search images of Boston Marathon bomber.
SOURCES: (A) Handout/Getty Images News via Getty Images, https://www.gettyimages.com/detail/news-photo/in-this-image-released-by-the-federal-bureau-of-news-photo/166984823. (B, C) Federal Bureau of Investigation, 2013, "News Surveillance Video Related to the Boston Bombings," https://www.fbi.gov/video-repository/newss-surveillance-video-related-to-boston-bombings.

12-million-individual mugshot database, all algorithms tested by NIST find the person in Figure 2-7(B) and correctly return the match—and 10 contemporary algorithms placed the correct image at rank 1. That occurs despite the blur, chin occlusion, and viewpoint change. In 2013, however, face recognition was not successful at identifying the perpetrators even though their photos were in governmental databases.[32] Even with a decade of improvements, none of the algorithms in 2023 succeeded at recognition using Figure 2-7(C) as the search photo owing to the blur, downward viewpoint, and shadow.

This result occurs despite ongoing research efforts focused on recognition of CCTV-captured and other images where neither the photographic environment nor the subject's viewpoint with respect to the camera are conducive to providing high-quality face images for recognition. When a face image simultaneously contains multiple confounding factors such as variations in facial pose, illumination, expression, occlusion, image resolution, and facial aging, facial recognition may succeed or fail depending on the extent of those problems. However, recognition performance degrades for unconstrained face images—where image acquisition is uncontrolled and subjects may be uncooperative—requiring human intervention for accurate recognition.

---

[32] S. Gallagher, 2013, "Why Facial Recognition Tech Failed in the Boston Bombing Manhunt," *Ars Technica*, updated May 7, https://arstechnica.com/information-technology/2013/05/why-facial-recognition-tech-failed-in-the-boston-bombing-manhunt.

ACCURACY

Face recognition works by comparing faces appearing in photos and producing measures of similarity. In most applications, a decision must be produced—for example, should the phone unlock, should the door open, or should a person board an aircraft without that person's identity document being checked? As with other biometric traits such as fingerprints, decisions are made by comparing the similarity score to a threshold. The threshold is set by the system owner, often based on a provider recommendation. The appropriate threshold (and the acceptable error rate) for a particular application depends heavily on the statistics of the images and the relative costs of false negative (FN) and FP matches for the application. In a one-to-one authentication context, the threshold is typically set so that it is unlikely that unauthorized access will be granted.

However, face recognition, as with other kinds of authentication, sometimes fails. The next sections give terms and definitions to the sorts of errors that occur. More formal and extensive definitions and requirements for testing of biometric systems can be found in the ISO/IEC 19795-1:2019 standard.[33]

Errors in One-to-One Verification Systems

Two types of error are possible. First is a *false negative match*, in which the face recognition algorithm fails to emit a similarity score above a decision threshold, and thereby fails to associate the two images of one face. Second is a *false positive match*, in which the algorithm produces a spuriously high score from images of *two* people.

A third category of error is possible: failures relating to cameras not collecting a photo (known variously as *failure to capture*, or *failure to acquire*) or of the algorithm failing to find or extract usable features from an image (*failure to enroll* or *failure to extract template*). Note that template generators can be configured to not produce an output if the input sample was of low quality; otherwise, a template may cause false matches or false non-matches in subsequent recognition. Quality assessment is considered essential to the ethical use of face recognition.

Errors in One-to-Many Identification Systems

One-to-many search systems take a photo of a face and return similar faces from one or more reference databases. For example, a person entering a casino could be searched against a database of known cheats, and against a database for high rollers. Face

---

[33] ISO, 2021, "Information Technology, Biometric Performance Testing and Reporting—Part 1: Principles and Framework," ISO/IEC 19795-1:2021.

recognition identification systems are generally configured in two ways—automated identification and investigational use.

Automated Identification

The system returns faces that are more similar than a numerical threshold. The threshold is specified by the system owner, and the users of the system must have a procedure to handle multiple matches.

With automated identification, FNs occur when the person in the probe image is present in the reference gallery but is not matched. FNs also occur because the algorithm finds the search photo to be dissimilar, at the specified threshold, to its reference gallery mate.

FPs occur when a non-mated search yields any candidates. A non-mated search is one in which the person in the photograph is not present in the reference gallery. FPs also occur when a comparison of the search photo and a reference gallery entry yields a similarity score at or above a threshold.

Investigational Use

The system is configured with a threshold of zero and returns the top K most-similar faces. The value K is usually specified by the system owner's policy. More rarely, the value might be set by the investigator running the search—for example, to lower the threshold in an investigation of a serious crime—and in a manner consistent with policy set by the system owner. In this configuration, human review is a necessary and integral part of what is then an automated-plus-human *system*.

FNs occur in mated searches when (1) the search does not include the correct mate in the top K candidates or (2) the search does place the correct mate in the candidate list, but the human reviewer misses it because they judge it to be a non-mate.

An FP can occur in two cases. The first case is a non-mated search where the human reviewer erroneously associates the search photo with one of the K candidate reference images. An FP can also occur for a mated search if the human reviewer misses the correct mate photo and instead associates the search photo with one of the other candidate reference images. Note, the FRT component returns K candidates whether the searched person is in the reference database or not, because the threshold is set to zero. This means that that the false positive identification rate (FPIR) of the FRT engine is 100 percent. If, instead, the algorithm were equipped with a somewhat higher threshold, candidate list lengths often would be reduced, thereby offering the human reviewer fewer opportunities to make an FP mistake, but also fewer opportunities to detect weakly matching mates.

The quality, thoroughness, and accuracy of the human review is critical to this process. In operational settings, the reviewer, who may not be an expert, or even trained, could be working under time-pressure or urgency imperatives related to the case. In such circumstances, mistakes will occur. Even without such exigencies, human review may not be reliable, as discussed later. The interaction between machine and human has previously been studied in the related area of latent fingerprint matching,[34] where a low-quality sample is compared with an exemplar print retrieved in a biometric search.

Primary Causes: What Typically Causes False Negatives and False Positives?

False Negatives

Face recognition is sensitive to changes in appearance of a subject. Consider the two photographs of musician John Lennon at different times in his life (Figure 2-8).

The primary causes of change in appearance are aging, poor photography, poor presentation, and acute injury. Poor photography reduces image quality, with typical manifestations being underexposure, overexposure, and misfocus. Poor presentation also reduces image quality, typically arising because the subject does not look at the camera, or moves, inducing motion blur. Many other factors can reduce pairwise similarity. These include occlusion (a waved hand or sunglasses, for example); resolution (face is too small or the camera's optics are poor); noise (owing to low light or weather); and image compression (owing to misconfiguration or low-bit-rate video).

In applications where subjects make cooperative presentations to a camera, FN rates can rise owing to poor usability. This is especially true in systems that are not used regularly—like border control gates—where subjects will not be habituated to the process. In such cases, usability testing is especially valuable. Some systems have good affordance and achieve low FN rates. Such systems usually allow a subject to retry.

False Positives

If a face recognition system erroneously associates photos of different people, an FP occurs. FPs arise primarily owing to similar appearance of two faces, which primarily arises from biological similarity of the faces, such as occurs in relatives, and particularly identical twins. This is discussed further in the discussion of demographic effects in the section "Demographic Disparities."

FPs can occur due to similarity of artifacts in images, such as similar thick-framed eyeglasses, or prominent nostrils. Such effects are idiosyncratic to the algorithm, and generally less common in recent algorithms.

In large-scale one-to-many identification systems, where tens or hundreds of millions of people could be represented in the reference database, there is an elevated

---

[34] I.E. Dror and J.L. Mnookin, 2010, "The Use of Technology in Human Expert Domains: Challenges and Risks Arising from the Use of Automated Fingerprint Identification Systems in Forensic Science," *Law, Probability and Risk* 9(1):47–67, https://doi.org/10.1093/lpr/mgp031.

FIGURE 2-8 Two photos of musician John Lennon at different times in his life illustrate change in appearance.
SOURCES: (*Left*) E. Koch, National Archives/Anefo, http://hdl.handle.net/10648/aa6be4d4-d0b4-102d-bcf8-003048976d84.
(*Right*) J. Evers, National Archives/Anefo, http://hdl.handle.net/10648/ab63fd72-d0b4-102d-bcf8-003048976d84.

chance of an FP match. In many systems, if the size of the reference database increases, the threshold will need to be increased to maintain a target FP identification rate. Some systems address this automatically.

Accuracy Improvements Over Time

The accuracy of a biometric system is estimated by conducting empirical trials. The result is a measurement of an FP rate and an FN rate. To compare systems, an analyst will configure a decision threshold for each system that yields a particular FP rate—say, 1 in 10,000—and then report the FN rate. Figure 2-9 shows how such a measure has improved since 2017 for algorithms from one industrial developer.

Figure 2-9 shows an analog of Moore's law with face recognition error rates reducing annually by approximately a factor of 2. This applies to three fixed databases, involving cooperative photographs from four operational sources. The FN rates reduce because the algorithms are increasingly able to associate poor-quality photos and those of faces taken up to 18 years apart. Although such gains have been realized by many developers, error rates vary considerably across the industry: some organizations produce algorithms that are much more accurate than others. Importantly, any given operator of face recognition can only realize such gains in its operations by procuring updated algorithms and applying them to its image databases. Another implication is that operators will find pairs of mated images in legacy databases that had previously been unknown; in a criminal justice investigation, this could produce a new lead.

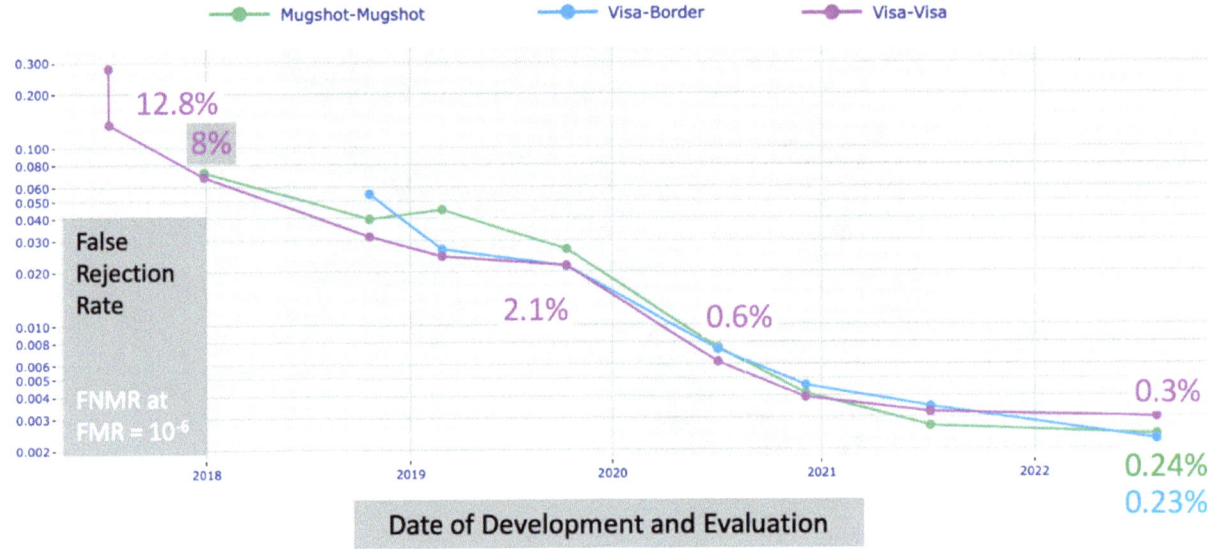

FIGURE 2-9 Date of development and evaluation versus false rejection rate for FRT from a single vendor.
SOURCE: National Institute of Standards and Technology, 2023, "Face Recognition Technology Evaluation (FRTE) 1:N Identification," Department of Commerce, https://pages.nist.gov/frvt/html/frvt1N.html.

Today, state-of-the-art systems are able to recognize images captured under controlled conditions with recognition accuracy high enough to meet many application requirements. They are also able to recognize poorer-quality photographs where the subject does not cooperatively engage the camera, or where the camera optics or imaging environment are poor. This ability has enabled end users to expand their capture envelope to include less-constrained photographs.

It is not possible to give a one-line answer to the question of how good *face recognition* is. Accuracy is inextricably linked to the properties of the images (both the search photo and database faces) being used. A second factor is the algorithm; accuracy varies widely across the industry. Face recognition algorithms do not yet have the capability to report "search photo does not exist in the database" without downgrading their capability to find true matches.

Accuracy in Large Populations

With an exception detailed below, face recognition search is viable even with databases with several hundred million faces—where viable means error rates that are sufficiently low for many use cases. First, the FP identification rate must be low—the system should not mismatch too many search photos with database entries—which is achieved by using a high threshold. However, the threshold cannot be raised arbitrarily because that will cause an elevation in FN identification rates—the system will fail to retrieve ("miss") matching database entries. There is thereby a trade-off between FN and FP error rates.

As more individuals are enrolled into a database, the possibility of a mismatch increases. To maintain a fixed FPIR, it is necessary either for the algorithm to adapt or for the system owner to raise the threshold. Search remains viable in very large populations because of an aspect of statistics concerned with tails of distributions. To limit the FPIR—the proportion of searches that return a mismatch when they should not—the algorithm must correctly report only low similarity scores. The highest score, known to statisticians as an *extreme value*, will grow as the number of people in the database grows. However, the highest value grows only slowly with the size of the database. By analogy, one will find taller people in a sample of 10,000 versus 1,000, but not that much taller.

However, there is a problem. The extreme value model implies that FPIR grows slowly so that thresholds need to be elevated only slightly to maintain FPIR. However, this assumes that the similarity scores are sampled from a single and stable population distribution—that is, that one does not expect outlier or freak scores. In the same way that 500-year floods will occur more frequently when the climate has changed, the actual non-mate distribution will include a well-known population that generates high non-mate scores: twins. Twins are common: 3 percent of newborns are a twin in the United States[35] and 0.4 percent are identical twins.[36] Twins are becoming increasingly common with later-in-life motherhood and increased use of fertility technologies. The effect on FRT is that if one twin is in the database, and the other is searched, an FP will occur (because contemporary FRT algorithms are incapable of distinguishing them). Such events occur naturally even in small populations—for example, if the entire population of a small town is enrolled. They will occur more frequently in large data sets such as state drivers' licenses databases.

Figure 2-10 shows that, even with the most accurate contemporary algorithms, low FPIRs cannot be achieved by elevating thresholds because FN rates ascend rapidly to levels that would render the system useless.

---

[35] Centers for Disease Control and Prevention, 2023, "Births: Final Data for 2021," *National Vital Statistics Reports* 72(1), https://www.cdc.gov/nchs/data/nvsr72/nvsr72-01.pdf.

[36] P. Gill, M.N. Lende, and J.W. Van Hook, "Twin Births," updated February 6, In *StatPearls [Internet]*, Treasure Island, FL: StatPearls Publishing, https://www.ncbi.nlm.nih.gov/books/NBK493200.

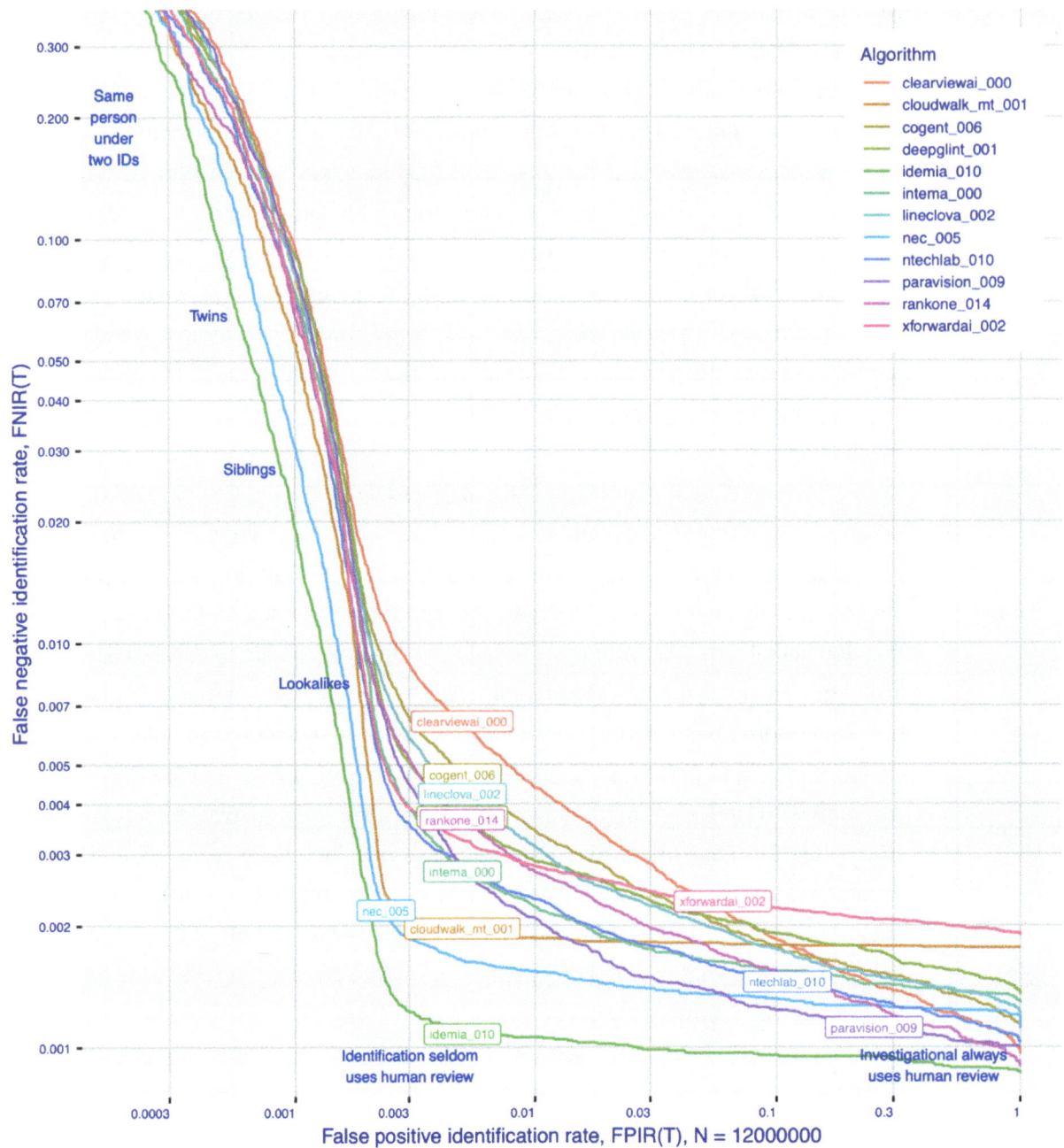

FIGURE 2-10 False positive identification rate of given algorithms.
SOURCE: National Institute of Standards and Technology, 2023, "Face Recognition Vendor Test (FRVT) Part 2: Identification," NISTIR 8271 Draft Supplement, Department of Commerce, https://github.com/usnistgov/frvt/blob/nist-pages/reports/1N/frvt_1N_report_2023_02_10.pdf.

DEMOGRAPHIC DISPARITIES

All machine learning–based systems, including biometric systems, potentially have performance that varies across demographic groups. (An analogous effect, the cross-race effect—that is, the tendency for individuals to more easily recognize faces that belong to their own racial group—is seen with human observers.) This arises fundamentally because humans vary anatomically: our characteristics differ individually, and by sex, by age, by ethnicity, and potentially other groupings that may not have descriptors associated with them. Some groups are categorical (e.g., sex), some are continuous (e.g., height), and some are defined as categorical (e.g., the young versus the old). It is the responsibility of a biometric system designer to ensure uniform function across all groups—or at least sufficiently close to uniform to be acceptable for a given application—or to qualify that the system should be augmented or not used by certain groups.

The first study of differential accuracy among different demographic groups was a 2003 report from NIST.[37] It found that female subjects were more difficult for algorithms to recognize than male subjects, and that young subjects were more difficult to recognize than older subjects.

Considerable attention has been paid to demographic effects in face recognition since the 2018 "Gender Shades" study of cloud-based algorithms that inspect a face image and return a classification of male or female.[38] The study showed that the algorithms tested misclassified the gender of women more than men, and those with dark skin tone more than light skin tone, and it gave the highest error rates on dark skin tone women, classifying up to 35 percent of African females as men. While the work had the effect of drawing attention to demographic performance differences in face recognition, the Gender Shades systems were not face recognition algorithms because they are not designed to support verification or determination of who a person is. Classification algorithms make a direct guess at gender. Recognition algorithms use different mechanisms—they encode identity into templates and, later, compare them. The persistent popular conflation of gender classification and face recognition may stem from the fact that algorithms used for both tasks employ neural networks trained on, respectively, large gender- and identity-labeled sets of photographs, although they are trained toward different objectives.

All face recognition system components potentially have error rates that depend on the demographics of the subjects. For example, a camera might have inadequate

---

[37] P.J. Phillips, P. Grother, R.J. Michaels, D.M. Blackburn, E. Tabassi, and M. Bone, 2003, *Face Recognition Vendor Test 2002: Evaluation Report*, NISTIR 6965, https://nvlpubs.nist.gov/nistpubs/Legacy/IR/nistir6965.pdf.
[38] J. Buolamwini and T. Gebru, 2018, "Gender Shades: Intersectional Accuracy Disparities in Commercial Gender Classification," *Proceedings of Machine Learning Research: Conference on Fairness, Accountability and Transparency* 81:1–15.

field of view to capture tall individuals; a face detector could fail on individuals with no hair and with eyebrows of similar color to their skin; a quality assessment algorithm might reject a passport application photo of an individual whose eyelids are very close to each other;[39] or a presentation attack detection algorithm might reject a face because it misclassifies long hair near the face as the edge of a device used to present a replay in image in a spoofing attack. For face recognition itself, both FP and FN error rates can differ. Importantly, the magnitudes, causes, and consequences of these errors differ, so they are discussed separately in the following two subsections. This separation adds specificity over statements made in many articles that face recognition does not work in a particular group.

The most thorough evaluation of disparity in face recognition across demographic groups was the 2019 NIST Face Recognition Vendor Test,[40] which raised awareness in the academic community and prompted vendors to collect additional training data and improve the facial recognition algorithm accuracy to reduce bias across the demographic groups.

False Positive Variation by Demographic Group

Nature. FPs involve two people: they occur when images of two people are incorrectly matched, which will occur when an algorithm returns a high similarity score. This can occur for a variety of reasons, depending on the algorithm. These include natural similarity of identical twins and other close relatives; spurious high scores from very poor-quality photographs such as low resolution or extreme overexposure; and matching within demographic groups that are under-represented in the data sets used to train the algorithm. FPs will also occur when the decision threshold is set to a very low value, as is the case when humans are employed to review the matches.

Affected groups. For most algorithms, FP rates are higher in women than men, also in the very young and old, and in particular ethnic groups.[41,42] For many algorithms, these groups are Africans, African Americans, East Asians, and South Asians. For some algorithms developed in China, the East Asian group gives low FP rates and, instead, the White group gives elevated rates. FPs are highest at the intersection of these groups—for

---

[39] J. Regan, 2016, "New Zealand Passport Robot Tells Applicant of Asian Descent to Open Eyes," *Reuters*, updated December 7, https://www.reuters.com/article/us-newzealand-passport-error/new-zealand-passport-robot-tells-applicant-of-asian-descent-to-open-eyes-idUSKBN13W0RL.

[40] P. Grother, M. Ngan, and K. Hanaoka, 2019, *Face Recognition Vendor Test (FRVT)—Part 3: Demographic Effects*, NISTIR 8280, Washington, DC: Department of Commerce and Gaithersburg, MD: National Institute of Standards and Technology, https://doi.org/10.6028/NIST.IR.8280.

[41] G. Pangelinan, K.S. Krishnapriya, V. Albiero, et al., 2023, "Exploring Causes of Demographic Variations in Face Recognition Accuracy," arXiv:2304.07175.

[42] K. Krishnapriya, V. Albiero, K. Vangara, M.C. King, and K.W. Bowyer, 2020, "Issues Related to Face Recognition Accuracy Varying Based on Race and Skin Tone," *IEEE Transactions on Technology and Society* 1(1):8–20.

example, for many algorithms elderly Chinese women give the highest false match rates. These effects are not related to poor photography; they occur even in well-controlled, standard-quality images. Also, this is not clearly related to skin tone—high false match rates are observed in both light-skinned East Asian and dark-skinned African populations. Furthermore, algorithms known to be trained on East Asians can give high false match rates on Whites. Last, very young children give high false match rates,[43] possibly due to undeveloped features and severe lack of representation in training sets.

Magnitude and prevalence. These will be more common in deployments where many non-mated comparisons are performed. This will occur in one-to-many searches of large databases such as when detecting duplicate identities in benefits systems, and when many non-mated searches are conducted—for example, in public area surveillance, or sports arena entry, where a watchlist alert system is in use. FP rates can vary massively across groups; the ratio can be one, two, or three orders of magnitude in some demographic groups versus others; this depends strongly on the algorithm and the groups being recognized.

Impact. The consequences of FPs vary by application. As an FP involves two people, either or both can be affected. In a one-to-one access control task, an FP could lead to loss of privacy or theft, for example. In a pharmacy, an employee would not be able to refute the assertion that they dispensed drugs to a fraudster. In a benefits-fraud detection setting, an FP might lead to a wrongly delayed or rejected application. In a public area surveillance application, an FP could result in interview and arrest.

Root-cause remediation. There is consensus that remediation of disparities in FP rates is the job of the recognition algorithm developer by, for example, increasing the diversity of the training data or accounting for imbalances in the training data by reweighting under-represented groups.[44]

False Negative Variation by Demographic Group

Nature. FNs involve one person: they occur when two photographs of that person do not match, which is a result of low similarity arising from some change in facial appearance. This can occur owing to a change in hairstyle or presence of cosmetics, to aging, or when image quality is degraded—for example, when a photograph does not have fidelity to a subject's face. This can occur variably across demographic groups. One common circumstance is for a photograph to be underexposed, a problem that occurs

---

[43] P.J. Grother, M. Ngan, and K. Hanaoka, 2019, *Face Recognition Vendor Test (FRVT)—Part 3: Demographic Effects*, NISTIR 8280, Washington, DC: Department of Commerce and Gaithersburg, MD: National Institute of Standards and Technology, https://doi.org/10.6028/NIST.IR.8280.

[44] M. Bruveris, J. Gietema, P. Mortazavian, and M. Mahadevan, 2020, "Reducing Geographic Performance Differentials for Face Recognition," Pp. 98–196, IEEE Winter Applications of Computer Vision Workshops (WACVW), https://doi.org/10.1109/wacvw50321.2020.9096930.

more frequently in dark-skinned individuals because pigmented skin reflects less light. Poor photography can lead to overexposure of light skin, but this is less common. Given such images, a face detector can fail such that a test might record failure-to-capture rates that differ by demographic group. If detection succeeds, however, an underexposed face image can have insufficient detail to allow the face recognition algorithm to discern face features or face shape. This will tend to elevate FN rates.[45]

Affected groups. Although FNs are usually more common in women than men, and sometimes in Africans and African Americans versus Whites, false FNs are uniformly quite low (see the following), and variation across groups is small. Standardized measures of inequity are much smaller than for FPs. An exception to this is in very young children, where rapid, growth-related changes in appearance cause FN rates to be much higher than in adults.

Magnitude and prevalence. Notably, with contemporary face recognition algorithms applied to images collected from cooperative subjects, FN rates are below 1 percent, and much lower than the gender misclassification rates measured in Gender Shades—for example, 35 percent. FN rates and demographic differences will generally increase if imaging is less controlled, such as from a webcam installed in a taxi being operated at night.

Impact. The consequences of an FN vary by application. In a mobile-phone authentication context, an FN can be remedied by a retry or entering of a PIN. Without a secondary authentication mechanism, a set of FNs in a time-and-attendance application could be construed as a failure to come to work. In a surveillance application, FNs are to the advantage of the person; in a protest, for example, an individual might wear a protective face mask and sunglasses to hide their features and thereby impede detection or induce an FN. Likewise, an FN would be to the benefit of a soccer hooligan.

The magnitude of demographic variation depends on what measures have been taken to mitigate these issues. For example, some systems use improved lighting to help mitigate face detection and insufficient detail effects with dark skin tone individuals. Some systems have attempted to rebalance the composition of the training data to mitigate the effects of under-representation.

Root-cause remediation. This is a photography problem that is difficult to fully remedy without adoption of controlled light, controlled exposure, high-dynamic-range imaging, or active camera control. The value of such approaches will be realized only if higher-precision data transmission standards[46] are promulgated in the face recognition

---

[45] C.M. Cook, J.J. Howard, Y.B. Sirotin, J.L. Tipton, and A.R. Vemury, 2019, "Demographic Effects in Facial Recognition and Their Dependence on Image Acquisition: An Evaluation of Eleven Commercial Systems," *IEEE Transactions on Biometrics, Behavior, and Identity Science* 1(1):32–41.

[46] ISO, 2022, "Information Technology-JPEG XL Image Coding System—Part 1: Core Coding System," ISO/IEC 18181-1:2022.

community; these would encode luminance (and color) in more than 8-bit integers, allowing higher-contrast images to be captured.

The impacts of errors associated with demographic variation depend on the application. For example, in authentication scenarios like access control, where almost all usage is by the legitimate account holder, a high FN rate in a demographic would directly impact convenience and useability. The same system will be configured to give low FP rates (1 in 10,000 is typical), such that even if some demographic existed for which the false match rate was much higher (1 in 100, say), then it would still be rare for there to be any observable impact. Indeed, some practitioners incorrectly consider FP variations to be entirely irrelevant, arguing that it only affects impostors. However, a high FP match rate can represent a security flaw such that members of an affected demographic could be harmed. In one-to-many surveillance applications, such as soccer stadium entry, FPs cause adverse outcomes (e.g., eviction), so large demographic variations are hazardous.

FACE RECOGNITION UNDER ATTACK

Face recognition is used to verify identity claims and to identify subjects in a database. In applications that are used to confer some benefit—such as access to a building, country, or account—a bad actor may seek to subvert the intended operation of the system. Depending on the setting, an attacker may want to positively match someone else, or to not match themselves. These are discussed in the next two subsections.

Impersonation

In verification, if an attacker can successfully use a face recognition system to match a victim, then the benefits accrue to the attacker—this could be access to a mobile phone, or entry to a country using someone's passport. This standardized term for this is impersonation,[47] and it requires the attacker to (1) appropriate a credential (the phone or passport), and (2) arrange for the face recognition to produce a sufficiently high similarity score. This is attempted in the physical domain using a number of techniques such as wearing a face mask or cosmetics so as to resemble the legitimate enrollee, or by simply displaying a photo of that person on paper or tablet. Such methods are termed presentation attack instruments, and the activity is a presentation attack. Examples are shown in Figure 2-11. It is also possible to launch attacks in the digital domain by injecting a photo electronically into a system—for example, by tricking the receiving system into thinking that the injected photo came from a real camera.

---

[47] ISO, 2023, "Biometric Presentation Attack Detection—Part 1: Framework" ISO/IEC 30107-1:2023.

 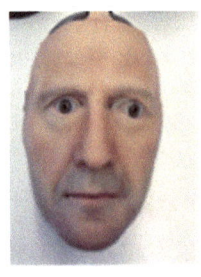 

Bona fide passport-style photo | 3D mask attack instrument | 2D printout replay attack instrument

FIGURE 2-11 Legitimate photo of a subject, and two presentation attack instruments.
SOURCES: P. Grother, M. Ngan, and K. Hanaoka, 2019, *Face Recognition Vendor Test (FRVT), Part 3: Demographic Effects*, NISTIR 8280, Washington, DC: National Institute of Standards and Technology, Department of Commerce, https://nvlpubs.nist.gov/nistpubs/ir/2019/NIST.IR.8280.pdf.

The success of such attacks depends on knowledge, opportunity, skill, and whether countermeasures, if any, are effective. The attacker generally needs to know who they are attacking—to impersonate the owner of a phone, an attacker will need knowledge of their appearance. This is often readily available via casual observation and photography of the victim. For other biometrics such as fingerprint or iris, such information is more difficult to come by.

Impersonation attacks are possible also in face recognition applications using one-to-many search. For example, in a paperless aircraft boarding application, a subject resembling someone on the departure manifest could authenticate and board successfully. An identical twin or an able attacker equipped with a face mask could attempt this. Such systems are single-factor authentication systems relying solely on the biometric match.

By using a presentation attack instrument that resembles a target subject, an impersonator could incriminate that person at a crime scene that they knew was being recorded.

Evasion

Face recognition is often used to check whether a subject has been seen previously. For example, if people are evicted from a casino for cheating, their photos may be retained and enrolled in a face recognition system with the intent that they will be recognized and denied entry should they return. An attacker would anticipate such steps and seek to evade recognition. This may be achieved by avoiding cameras, by not looking at cameras, or, more effectively, by changing one's appearance so that recognition returns a low similarity score. This can be attempted by wearing a face mask of someone else, by wearing sufficient cosmetics, or by occlusion. For example, in the 2019 protests against legislation in Hong Kong, citizens wore face masks to undermine recognition.

Detection of Attacks

Attack detection is critical in applications where economic or other incentives exist for attackers to impersonate or evade. For example, there are obvious monetary benefits to someone who can execute unemployment benefits fraud by establishing two or more identities. As such, there are successful efforts to detect presentation and injection attacks. These fall into two categories: passive and active. PAD analyzes the received biometric data, which could be a photo or video, and makes a decision. In active attack detection, the software will arrange for a change in the appearance of a subject—for example, by issuing an instruction to the subject, or by manipulating the illumination of the subject. The key to success of such countermeasures is randomness: the attacker would need to respond correctly to the "challenge" issued by the PAD system. Both passive and active attack detection schemes can be supplemented with information obtained from other sensors—for example, the vascular structure of a face could be imaged using a long-wave infrared camera sensitive to thermal information.

If attack detection can be done perfectly, then the biometric system conclusively binds the actual person to the capture event. If it is imperfect, then security and trust are eroded.

HUMAN ROLES AND CAPABILITIES

In applications of face recognition such as access control, where most transactions are mated, accuracy is high enough that matching will usually succeed. In those FN cases where it does not, a secondary resolution process is needed. This could involve a human, as happens after a passport gate rejection in immigration, or with an airline staff member after a failure in automated aircraft boarding. In such cases, the human will compare the face on a presented ID document with that of the identity claimant. This process will itself have some errors: FNs if the reviewer fails to verify a legitimate claimant and FPs if an impostor is verified—for example, when the impostor is trying to circumvent the automated check.

In investigations, face recognition is typically used to present lists of candidate photos to a human reviewer, who compares each candidate with the searched photo to check whether it is a true match. The use of human review is an integral part of the process, used in 100 percent of searches. Moreover, humans are fallible and, as with FRT, human review can result in two types of error. These are FPs (incorrect associations of two people in the photos) and FNs (failure to associate one person in two photos). In criminal investigations, an FN would result in an unidentified suspect, but an FP could lead to an incorrect detention. When humans review long lists of candidate photos, there are typically tens of opportunities for false matches: the human review must correctly reject *all* of them to avoid an FP. In terms of binomial statistics, even if a reviewer's false match rate was 1 percent, then the chance of falsely accepting any one of 50 would be $1 - (1 - 0.01)^{50}$—which is about 0.4, or about a 40 percent chance that a mistake will be made.

Human adjudication of photos has been extensively studied by experimental psychologists. The task is termed "unfamiliar face matching," as it usually involves review of two juxtaposed photos to determine whether they are of the same person. As such, the task does not require memorization. The first step for a human is to determine if one or both of the photos are unsuitable for comparison; this "no value" determination is sometimes skipped, and a match or no-match decision will be made. Face recognition algorithms faced with the same task can fail to find a face or can electively refuse to process an image by analyzing its quality and suitability for recognition. However, systems are usually configured to accept even poor-quality photos.

It is well documented that a human reviewer's accuracy is improved when there are no constraints on review duration,[48,49] there are multiple images of a person,[50] the images are of standardized high quality,[51,52] and the reviewer has had adequate sleep.[53] Additionally, it is known that accuracy depends on the demographics of the reviewed faces—most importantly, that humans of one race give reduced accuracy when reviewing photographs of another.[54] Human false non-match rates are reduced when the expression and head orientation in the two photos are similar and when the time elapsed between photo creation is small.[55]

Human trials are complicated because human performance is time dependent on timescales similar to the test duration, and over longer timescales. One notable aspect

---

[48] M.C. Fysh and M. Bindemann, 2017, "Effects of Time Pressure and Time Passage on Face-Matching Accuracy," *Royal Society Open Science* 4(6).

[49] M. Özbek and M. Bindemann, 2011, "Exploring the Time Course of Face Matching: Temporal Constraints Impair Unfamiliar Face Identification Under Temporally Unconstrained Viewing," *Vision Research* 51(19):2145–2155.

[50] D. White, A.M. Burton, R. Jenkins, and R.I. Kemp, 2014, "Redesigning Photo-ID to Improve Unfamiliar Face Matching Performance," *Journal of Experimental Psychology: Applied* 20(2):166.

[51] A.M. Burton, D. White, and A. McNeill, 2010, "The Glasgow Face Matching Test," *Behavior Research Methods* 42(1):286–291.

[52] P.J. Phillips, 2017, "A Cross Benchmark Assessment of a Deep Convolutional Neural Network for Face Recognition," Pp. 705–710 in *2017 12th IEEE International Conference on Automatic Face & Gesture Recognition (FG 2017)*, Washington, DC, https://doi.org/10.1109/fg.2017.89.

[53] L. Beattie, D. Walsh, J. McLaren, S.M. Biello, and D. White, 2016, "Perceptual Impairment in Face Identification with Poor Sleep," *Royal Society Open Science* 3(10):160321.

[54] C.A. Meissner and J.C. Brigham, 2001, "Thirty Years of Investigating the Own-Race Bias in Memory for Faces: A Meta-Analytic Review," *Psychology, Public Policy, and Law* 7(1):3–35.

[55] A.M. Megreya, A. Sandford, and A.M. Burton, 2013, "Matching Face Images Taken on the Same Day or Months Apart: The Limitations of Photo ID," *Applied Cognitive Psychology* 27(6):700–706.

is that human observers gradually develop a match bias during prolonged testing such that the FN rate declines (i.e., improves) but the false match rate increases.[56] Such behavior would be important, for example, over the hours of a border guard's shift. It would be less important in a criminal investigation featuring ample review time, and limited numbers of image pairs to review.

The cognitive explanation for the experimental observations is still being researched, but the existence of the effects, and their magnitudes, is largely settled. An important topic in cognition research is whether standardized forensic-level training is effective in improving accuracy. As an explanation, it has been suggested that training drives toward an unlearning of the innate perceptual mode in which humans process faces holistically.[57]

So how accurate are humans? In a 2017 test of human capability, reviewers were given three months to review 20 pairs of frontal photographs without being given identity ground truth; there were 12 pairs of the same person, and 8 pairs of different people.[58] The reviewers were categorized into five groups by experience, training, and aptitude: forensic examiners (with extensive training, and who testify in court); reviewers (who typically perform initial law enforcement reviews in investigations); super recognizers (who have documented aptitude in tests or during employment); and fingerprint examiners and undergraduate students (as control groups). Despite the extended review duration, only 7 of 57 examiners correctly adjudicated all 20 pairs. The corresponding figure for reviewers was 2 of 30, for super recognizers 3 of 13, for fingerprint examiners 1 of 53, and for students 0 of 31. More tangibly, for the most proficient groups, forensic face examiners and super recognizers, the study estimated an approximately 1 percent probability of assigning a highly confident match decision to an actually non-matching pair. The study did not address image quality. The images used were of fair quality, collected in a cooperative university setting.

OTHER SALIENT ATTRIBUTES OF TODAY'S COMMERCIAL FACIAL RECOGNITION TECHNOLOGY

Today's commercial FRT systems have several attributes that relate to how they might best be governed. These include

---

[56] H.M. Alenezi and M. Bindemann, 2013, "The Effect of Feedback on Face-matching Accuracy," *Applied Cognitive Psychology* 27(6):735–753.

[57] D. White, A. Towler, and R.I. Kemp, 2021, "Understanding Professional Expertise in Unfamiliar Face Matching," *Forensic Face Matching* 62–88.

[58] P.J. Phillips, A.N. Yates, Y. Hu, et al., 2018, "Face Recognition Accuracy of Forensic Examiners, Superrecognizers, and Face Recognition Algorithms," *Proceedings of the National Academy of Sciences* 115(24):6171–6176.

- *Proprietary.* Since its inception, the face recognition industry is built on algorithms that are trade secrets—the details of their architecture, objective functions, and training data are closely held. There are a few open-source algorithms, and although these may seed commercial development, they are not supported and documented to the level of commercial viability.
- *Not commoditized.* Commercial FRT algorithms vary greatly in their technical capabilities, in terms of accuracy, stability across demographic groups and imaging conditions, and in speed, memory, and power consumption. They differ also in the software maturity, application programming interface support for programmers, scalability to large populations and volumes of searches, and portability across computer hardware.
- *Deployed as cloud services as well as on-premises.* For many years, face recognition systems were deployed only as software libraries installed on customer-owned computers or cameras. In recent years, with widely deployed fast networks, face recognition systems have been deployed in clouds in which imagery is uploaded to a remote data center. The two deployment paradigms differ with respect to custody of customer data. In the on-premises approach, faces and associated biographic data are maintained on customer-controlled systems. In the cloud-based arena, the data are uploaded to cloud provider's hardware. As such, use of the data by the cloud provider is constrained only by the contractual arrangements between the cloud provider and the customer. Developers of cloud-based face recognition can train on customer data sets if they are not contractually barred from doing so, and if the images are accompanied by ID labels.

3

Use Cases

Facial recognition technology (FRT) is increasingly widespread, with use cases ranging from unlocking smartphones and other devices to uses in law enforcement investigations, at international borders, in airports, and in many other public and private spaces. FRT has become embedded in many aspects of everyday life, and it is expected that it will find its way into an increasing number of applications in the future.

This chapter describes a large range of FRT use cases and public discourse around these uses. Many use cases may be valuable and worthwhile—although there may be debate about the cases where FRT use is most appropriate and cost-effective. Many FRT use cases raise significant questions related to fairness, equity, civil liberties, or privacy, and the reader may find some use cases to be problematic. This is intentional. The committee deliberately included use cases ranging from the relatively innocuous and widely (albeit not universally) accepted to use cases that many believe should be prohibited.

The chapter deliberately does not address the normative dimensions of these use cases. Chapter 4 broadly considers equity, privacy, and civil liberties implications of these and other uses, and the conclusions and recommendations in Chapter 5 are informed by these use cases.

The chapter is divided into sections describing broad categories of use. Within each section, examples of use cases are presented that are currently deployed in the United States or internationally. Technically feasible—but currently hypothetical—use cases are also described. Recommendations for mitigating the more concerning issues associated with the use of FRT and a framework for assessing various use cases are discussed in Chapter 5.

LAW ENFORCEMENT INVESTIGATION

The use of security camera footage to identify suspects in a criminal investigation is one of the most common applications of FRT. Law enforcement agencies frequently seek to identify individuals from images captured using public or private video cameras. A law enforcement use case is provided below.

Law enforcement identification of a suspect from photo (current use): Police have a photo of a suspect fleeing the scene of a robbery. The photo is used to search a database of mugshots or area parolees, and one person is identified as a likely match.[1] Officers are then sent to question that person.

FRT can be applied to conventional security camera footage long used by many businesses and police investigations. The falling cost of high-quality cameras, network infrastructure, and storage has led to widespread surveillance in public and commercial spaces. The personal use of cameras, such as in doorbell systems and smartphones, has increased dramatically in recent years. Although they provide footage of varying quality, many of these cameras can capture images of sufficient quality for FRT. Public, commercial, and private video footage is commonly accessible by law enforcement investigators, and some police departments have developed formal programs to access private cameras.

PUBLIC SAFETY

FRT can potentially be used in high-traffic areas and during large gatherings, ranging from concerts or music festivals, parades, and sporting events, to social and political demonstrations, all of which are settings in which monitoring with FRT may be of interest to law enforcement and national security agencies—and which may in some cases raise civil liberties concerns. The use of FRT presents law enforcement with enhanced capacity to surveil large crowds to develop intelligence, detect persons previously identified as security risks, and stop a potential threat to safety.

Screening entrants to a concert against a list of known threats (current use): An arena uses FRT at the entrance and throughout the arenas and stadiums, scanning ticketed attendees against a list of individuals who pose specific threats to the performer. If

---

[1] Note: There may not be any likely match if the suspect is not in the database.

the FRT signals a likely match, the individual is questioned by venue security personnel and asked to show identification (ID). If the ID shows that the person is on the list, they may be barred from entry.

Over the past 10 years, the use of FRT in sports venue security has become commonplace.[2] The metal detector systems for entry into these venues are being replaced by "smart" entry gates. The systems increase flow of traffic into the venues and have had the benefit of protecting entertainers from their stalkers.[3] As an example, the systems used for Taylor Swift concerts have been effective in keeping the artist safe from known stalkers. Her security detail estimates that she has approximately 3,000 known stalkers, many of whom attend her shows.

Screening for shoplifters in stores (current use): A grocery store uses FRT at customer entrances, seeking to identify known shoplifters and deny them entry. The list of known shoplifters is compiled jointly by the store's security team and the security teams of competing local stores of individuals who have previous shoplifting offenses.

Increasingly, many major retail store chains are using facial recognition for security purposes.[4] Both shoplifting and "smash and grab" incidents have led some retailers to elect to deploy the systems, although sometimes in a limited fashion. In most instances, stores that have been the victims of repeated incidents have deployed FRT systems to assist law enforcement and to deter criminals. A major issue has been the disposition of the face images after they are collected; practices related to how long images are kept and with whom they are shared vary. There have been lawsuits filed on these grounds against certain chains to stop the deployment of FRT.[5] Another concern is how this use might have racial or other discriminatory effects on access to de facto public spaces.

Identifying card-counters and cheaters at casinos (current use): Casinos share lists of individuals banned from the premises for suspected card counting and cheating. Cameras are used to capture images of individuals entering casinos, and FRT is used

---

[2] *ABC News*, 2001, "Biometrics Used to Detect Criminals at Super Bowl," *ABC News*, updated February 13, https://abcnews.go.com/Technology/story?id=98871.

[3] B. Reed, ed., 2023, "Police to Use Live Facial Recognition in Cardiff During Beyoncé Concert," *Guardian*, updated May 17, https://www.theguardian.com/technology/2023/may/17/police-to-use-facial-recognition-technology-in-cardiff-during-beyonce-concert.

[4] J. Formoso, 2023, "Stores Are Using Facial Recognition to Help Stop Repeat Shoplifters," *FOX 5*, New York, updated March 17, https://www.fox5ny.com/news/stores-are-using-facial-recognition-to-help-stop-repeat-shoplifters.

[5] D.A. Ryskamp, 2021, "Macy's Faces Lawsuit Over Clearview AI Facial Recognition Software," Expert Institute, updated April 5, https://www.expertinstitute.com/resources/insights/macys-faces-lawsuit-over-clearview-ai-facial-recognition-software.

to compare these faces against those of known card counters or cheats. When a match is identified, the casino dispatches security to remove the individual from the facility.

Many casinos along the Las Vegas strip, and elsewhere in the United States, have implemented FRT to supplement security and monitor prohibited activity, allowing for real-time identification of individuals who were previously barred from gaming establishments to be blocked or removed.[6,7,8] Relatedly, individuals with a gambling addiction can choose to voluntarily enroll themselves in a database of known addicts; when individuals in this database show up to a casino and are identified using FRT, security personnel will similarly remove them from the casino.

School security—for example, identifying adults known to be dangerous (current use):
A public school has a list of adults known to be dangerous, such as individuals convicted of violent crimes. When visitors enter school property, their faces are compared against those of individuals on the list. If the system identifies a match, school security officers are immediately dispatched to escort the individual from school property.

In response to the recent school shootings in the United States, several school systems have deployed FRT on school grounds and in school buildings to monitor for unwanted persons on campus or in the building.[9,10,11] Goals include identifying bad actors, such as violent ex-students, registered sex offenders, non-custodial parents, and others deemed credible threats by law enforcement and school authorities. Parents and guardians of individuals enrolled in private schools would be able to consent to the use of FRT for broader use in surveillance of students, parents, staff, and visitors.[12] In this

---

[6] *Journal Record* Staff, 2022, "Casino Uses Facial Recognition Technology to Supplement Security," *Journal Record*, updated October 26, https://journalrecord.com/2022/10/casino-uses-facial-recognition-technology-to-supplement-security.

[7] C. Swanger, 2021, "How Integrated Resorts and Casinos Are Leveraging Facial Recognition Software for Increased Security," *eConnect Global*, updated November 19, https://www.econnectglobal.com/blog/how-integrated-resorts-and-casinos-are-leveraging-facial-recognition-software-for-increased-security.

[8] T. Prince, 2018, "Facial Recognition Technology Coming to Las Vegas Strip Casinos," *Las Vegas Review Journal*, October 13, https://www.reviewjournal.com/business/casinos-gaming/facial-recognition-technology-coming-to-las-vegas-strip-casinos.

[9] RealNetworks, 2018, "RealNetworks Provides SAFR Facial Recognition Solution for Free to Every K-12 School in the U.S. and Canada," updated July 17, https://realnetworks.com/press/releases/2018/realnetworks-provides-safr-facial-recognition-solution-free-every-k-12-school-us.

[10] D. Alba, 2020, "Facial Recognition Moves into a New Front: Schools," *New York Times*, February 6, https://www.nytimes.com/2020/02/06/business/facial-recognition-schools.html.

[11] C. Schulz, 2023, "Four Counties to Implement Facial Recognition for School Safety," West Virginia Public Broadcasting, https://wvpublic.org/four-counties-to-implement-facial-recognition-for-school-safety.

[12] SAFR® RealNetworks, n.d., "Leading by Example: How St. Therese Turned to SAFR® to Better Protect Its Community, One Opt-In at a Time," Case Study: Schools & Universities, https://safr.com/case-studies/st-therese, accessed May 23, 2023.

instance, a database of all staff and students who are approved for regular entry into the school or parents, guardians, or other visitors approved to pick up students may be created to verify their identity upon entry.

Other video analytics systems have been developed to identify a person approaching the school building with a weapon and alerts are sent to the school's access control system to trigger lock-down procedures. Although these video systems also have the capability for facial recognition, many school administrators are not using this capability with these systems due to privacy issues around FRT.

> *Human trafficking detection (current use):* Law enforcement agencies share information on persons reported as missing. At major transportation hubs such as airports, train stations, and ports, cameras are used to capture images of travelers. These images are compared against the shared database of missing individuals. When a match is identified, law enforcement is notified and dispatched to the transportation hub.

An important application of surveillance using FRT is to deter and detect the trafficking of humans, including the tracking of abducted children. Different types of trafficking observed in all states and territories of the United States include the movement of individuals and children for forced labor purposes and the sex trafficking of individuals, including young boys and girls. The National Child Protection Task Force[13,14] claims to use FRT as part of its enforcement arsenal that also includes geolocation and cellular data analysis. FRT has also been used to search online sex ads to find images matching those of a missing person.[15,16] Unfortunately, the exact impact of FRT systems on human trafficking has not yet been measured. An important future direction for FRT systems in humanitarian applications would be to measure the technology's impact.

> *Automated detection of offenses and offenders (current use outside the United States):* A police department places cameras in public areas. They are able to monitor the footage and identify infractions such as littering and can use FRT on the video footage to identify the culprits. The police are then able to generate and send citations for these infractions without needing additional officers on the street.

---

[13] M. Bernhard, 2021, "How NCPTF Helps Law Enforcement Find Missing Children," Skopenow, https://www.skopenow.com/news/how-ncptf-helps-law-enforcement-find-missing-children.

[14] T. Simonite, 2019, "How Facial Recognition is Fighting Child Sex Trafficking," *Wired*, https://www.wired.com/story/how-facial-recognition-fighting-child-sex-trafficking.

[15] B. Eastman, 2021, "Can Facial Recognition Software Within Transportation Technology Combat Modern Slavery and Human Trafficking?" *Futurist Journal of Law and Mobility*, https://futurist.law.umich.edu/can-facial-recognition-software-within-transportation-technology-combat-modern-slavery-and-human-trafficking/l.

[16] Department of Defense, 2017, "DARPA Program Helps to Fight Human Trafficking," https://www.defense.gov/News/News-Stories/Article/Article/1041509/darpa-program-helps-to-fight-human-trafficking.

Some countries with authoritarian regimes have deployed FRTs for automated detection of offenses and have most of their citizens in a database.[17,18,19] Another possible use would be to extend red light camera enforcement, which currently is based on license plate recognition, by using FRT to identify the driver and not just the car.

> *Identification as part of a traffic or street stop (hypothetical):* A police officer conducts a traffic stop or a stop of a pedestrian, acting on reasonable suspicion that the individual may have committed a crime. The officer takes a photo of the individual's face using a mobile device and accesses FRT to match the individual against a database of, for example, driver's license images, to establish the individual's identity.

Despite lack of reports of FRT use as a part of a stop by police, this hypothetical use case was recommended in early 2021 by Street Cop Training, a popular workshop on new investigative techniques.[20,21] In this case, police officers could use FRT to identify drivers or passengers, if their identity is uncertain, and determine whether the individual has a warrant for their arrest.

> *Anticipatory surveillance of crowds at a political protest (current use):* A police department fears that a large protest may become violent. The police use FRT to scan the crowd for matches against a list of known violent offenders and use this information to focus their attention and resources.

Large gatherings, such as social or political demonstrations, and public parades or celebrations, pose unique challenges for law enforcement to ensure the safety and security of both bystanders and protestors exercising their First Amendment rights. The use of FRT presents law enforcement and national security agencies with enhanced capacity to be able to surveil large crowds and potentially detect persons previously identified as posing security risks. There are potential risks associated with this use as well. As an example, questions have been raised as to whether it was appropriate to

---

[17] A. Ng, 2020, "How China Uses Facial Recognition to Control Human Behavior," *CNET*, https://www.cnet.com/news/politics/in-china-facial-recognition-public-shaming-and-control-go-hand-in-hand.

[18] K. Johnson, 2023, "Iran Says Face Recognition Will ID Women Breaking Hijab Laws," *Wired*, https://www.wired.com/story/iran-says-face-recognition-will-id-women-breaking-hijab-laws.

[19] *CBS News*, 2019, "Reporter on China's Treatment of Uighur Muslims: 'This Is Absolute Orwellian Style Surveillance,'" *CBS News*, https://www.cbsnews.com/news/china-puts-uighurs-uyghyrs-muslim-children-in-prison-re-education-internment-camps-vice-news.

[20] M. DeGeurin, 2022, "What to Do If a Cop Tries to Scan Your Face During a Traffic Stop," *Gizmodo*, https://gizmodo.com/can-police-use-facial-recognition-scans-at-traffic-stop-1848581619.

[21] C. Haskins, 2022, "A Popular Workshop for Police Encouraged Cops to Use Face Scans to ID People They Pull Over at Traffic Stops," *Business Insider*, https://www.businessinsider.com/police-workshop-street-cop-training-podcast-facial-recognition-traffic-stops-2022-2.

use FRT to identify Black Lives Matter protesters in Baltimore and New York.[22,23] In 2015, police in Baltimore County, Maryland, used facial recognition on photos retrieved from social media to identify individuals with outstanding warrants in the wake of events that transpired after the death of Freddie Gray.[24]

> *Scanning passersby in public places for outstanding arrest warrants (current use outside the United States):* City law enforcement maintains a list of individuals with outstanding warrants. A series of city-owned cameras in public spaces capture images of passersby. By using FRT to compare these images with the images of individuals with outstanding warrants, law enforcement can identify the whereabouts of these individuals and arrest them pursuant to the warrant.
>
> *Screen for parolees at travel sites (hypothetical):* A state parole agency puts FRT in local airports, bus stations, and car rental offices, looking for parolees whose conditions of parole forbid them from traveling outside the state. If there is a match, the facility is instructed not to offer travel to the person unless the person is cleared to travel.
>
> *Real-time mass surveillance (current use outside the United States):* The government sets up an extensive network of surveillance cameras across a city. Using FRT and the camera network, the government seeks to track the movement of any individual citizen to, for example, monitor criminal activity in high-crime areas or track movement of suspected terrorists.

Although the use of FRT video systems is not used in the United States as described in some of the above use cases, FRT could be deployed to screen specific areas for specific purposes or monitor individuals accessing public areas for undefined purposes where security camera infrastructure already exists. For example, an individual wanted by police was arrested after being identified by FRT at a concert from images collected by a network of closed-circuit television (CCTV) cameras around the public venue.[25,26] This

---

[22] Geofeedia and Baltimore County Police Department, "Case Study: Baltimore County PD," posted online by the ACLU of Northern California on October 11, 2016, https://www.aclunc.org/docs/20161011_geofeedia_baltimore_case_study.pdf.

[23] J. Vincent, 2020, "NYPD Used Facial Recognition to Track Down Black Lives Matter Activist," *Verge*, https://www.theverge.com/2020/8/18/21373316/nypd-facial-recognition-black-lives-matter-activist-derrick-ingram.

[24] R. Brandom, 2016, "Can Facebook and Twitter Stop Social Media Surveillance?" *Verge*, https://www.theverge.com/2016/10/12/13257080/police-surveillance-facebook-twitter-instagram-geofeedia.

[25] *BBC News*, 2018, "Chinese Man Caught by Facial Recognition at Pop Concert," *BBC News*, https://www.bbc.com/news/world-asia-china-43751276.

[26] A.B. Wang, 2021, "A Suspect Tried to Blend in with 60,000 Concertgoers, China's Facial-Recognition Cameras Caught Him," *Washington Post*, https://www.washingtonpost.com/news/worldviews/wp/2018/04/13/china-crime-facial-recognition-cameras-catch-suspect-at-concert-with-60000-people.

use case could be extended to other high-traffic areas, such as transportation hubs, to monitor individuals who are barred from leaving the state or country.

Some law enforcement agencies may also seek to use FRT technologies in an open-ended way to continuously scan passersby in public places, including public parks, streets, sidewalks, and public transportation centers, with no identified threat or concern.[27,28,29,30,31,32] The Metropolitan Police in the United Kingdom, for instance, announced in 2020 that they would begin to use live facial recognition in some public spaces to continuously scan for criminal suspects.[33] This deployment of live FRT is intended for use with a watchlist of wanted offenders or those persons who pose a risk of harm to themselves or others.[34] Examples of real-time mass surveillance in the United States have been linked to individuals sympathetic to foreign governments or government agents who have used existing CCTV footage to monitor individuals from their own nations residing in major U.S. cities, such as New York and Los Angeles.[35]

IN LIEU OF OTHER METHODS FOR VERIFYING IDENTITY OR CONFIRMING PRESENCE

Some applications of FRT are used to confirm identity by checking an individual's photo ID against a specific list of known exemplars. In these cases, deployment of FRT is intended to improve efficiency and provide enhanced security by allowing a search through multiple databases. It is important to note that there is a distinction among applications where facial recognition is utilized but individuals who prefer must specifically opt-out and where facial recognition is a convenience feature available for voluntary adoption.

---

[27] P. Mozur, M. Xiao, and J. Liu, 2022, "How China Polices the Future: An Unseen Cage of Surveillance," *New York Times*, p. A1, June 25.

[28] I. Qian, M. Xiao, P. Mozur, and A. Cardia, 2022, "China's Expanding Surveillance State," *New York Times*, p. A10, July 27.

[29] M. Xiao, P. Mozur, I. Qian, and A. Cardia, 2022, "China's Surveillance State Is Growing: These Documents Reveal How," *New York Times*, June 21, https://www.nytimes.com/video/world/asia/100000008314175/china-government-surveillance-data.html.

[30] P. Mozur, C. Fu, and A. Chien, 2022, "How China's Police Used Phones and Faces to Track Protesters," *New York Times*, updated December 4, https://www.nytimes.com/2022/12/02/business/china-protests-surveillance.html.

[31] D. Davies, 2021, "Facial Recognition and Beyond: Journalist Ventures Inside China's 'Surveillance State,'" *NPR*, https://www.npr.org/2021/01/05/953515627/facial-recognition-and-beyond-journalist-ventures-inside-chinassurveillance-sta.

[32] K. Hao, 2023, "After Feeding Explosion of Facial Recognition, China Moves to Rein It In," *Wall Street Journal*, https://www.wsj.com/articles/china-drafts-rules-for-facial-recognition-use-4953506e.

[33] A. Satariano, 2020, "London Police Are Taking Surveillance to a Whole New Level," *New York Times*, updated October 1, 2021, https://www.nytimes.com/2020/01/24/business/london-police-facial-recognition.html.

[34] Metropolitan Police United Kingdom, "Facial Recognition Technology: Live Facial Recognition," https://www.met.police.uk/advice/advice-and-information/fr/facial-recognition-technology, accessed November 17, 2023.

[35] I. Vincent, 2023, "After FBI Busts Chinese 'Police Station' in NYC, Six More Exposed in US," *New York Post*, updated April 19, https://nypost.com/2023/04/18/chinese-police-stations-allegedly-spying-on-nyc-la-more.

Border control for air travel (current use): An individual who is traveling internationally to the United States must go through customs and border control to confirm identity before entering the country. The traveler's photo is taken at a kiosk and compared against an existing passport or visa photo using FRT and confirmed within seconds. A Customs and Border Protection (CBP) officer can then interview the traveler and determine admissibility into the United States. For U.S. citizens, if the entry into the United States goes smoothly, the traveler's photo is deleted within 12 hours without further dissemination. For non-citizens, photos are retained for 14 days for facial comparison, then stored by the Department of Homeland Security (DHS) with entry and exit records.[36]

Over the past 20–30 years, the collection and processing of biometrics have become an important part of controlling movement at U.S. borders.[37] The ever-increasing numbers of these movements have motivated government agencies involved in the control of these movements to adopt biometrics to assist with these increasing numbers. The most widely used biometric for the identification and verification of persons crossing the border has been the analysis of fingerprints. However, CBP is replacing fingerprint identification with facial recognition to enable contactless and faster image acquisition and using it for comparison with face photos that are integrated in passports. CBP's Travel Verification System (TVS) compares a live photo of the traveler against a database of images from passports, U.S. visas, or other DHS holdings.[38]

To date, CBP has implemented FRT into entry processes at all international airports, into exit processes at 36 airport locations, and both entry and exit processes at 36 seaports, and all pedestrian lanes at both Northern and Southwest Border ports of entry. To date, CBP reports that it has processed more than 300 million travelers using biometric facial comparison technology and prevented more than 1,800 "impostors" (i.e., individuals using genuine travel documents that do not match their identity) from entry to the United States.[39] Currently at 18 seaports across the United States, CBP has partnered with Carnival Cruise Line[40] and Norwegian Cruise Line[41] to implement facial biometrics

---

[36] R. Iyengar and C. Gutman-Argemí, 2023, "How Technology Is Changing Immigration Lines," *Foreign Policy*, https://foreignpolicy.com/2023/04/27/us-immigration-lines-cbp-facial-recognition.

[37] Department of Homeland Security (DHS) and Customs and Border Protection (CBP), 2017, *Biometric Entry-Exit Program Concept Operations*, Washington, DC: Department of Homeland Security, https://epic.org/wp-content/uploads/foia/dhs/cbp/biometric-entry-exit/Concept-of-Operations.pdf.

[38] CBP, 2022, "Statement for the Record on Assessing CBP's Use of Facial Recognition Technology," updated August 29, https://www.cbp.gov/about/congressional-resources/testimony/statement-record-assessing-cbps-use-facial-recognition-technology.

[39] CBP, 2023, "Biometrics," updated October 5, https://www.cbp.gov/travel/biometrics.

[40] CBP, 2023, "CBP, Carnival Cruise Line Introduces Facial Biometrics at Port of Jacksonville," updated March 15, https://www.cbp.gov/newsroom/national-media-release/cbp-carnival-cruise-line-introduces-facial-biometrics-port.

[41] CBP, 2023, "CBP and Norwegian Cruise Line Introduces Facial Biometrics at the Port of Boston," updated April 10, https://www.cbp.gov/newsroom/local-media-release/cbp-and-norwegian-cruise-line-introduces-facial-biometrics-port-boston.

to further secure and streamline the identity verification process when travelers depart a vessel after a closed-loop cruise, reducing debarkation times by up to 30 percent.

For domestic travel, the Transportation Security Administration (TSA) has created a similar facial recognition program that conducts one-to-one matching compared to the photograph on their ID. TSA has implemented this pilot FRT program as part of the Touchless Identity Solution for PreCheck holders at 25 airports nationwide.[42,43] TSA is preparing to expand this program to more than 400 airports in the next several years[44] and is also testing a one-to-many facial recognition program where a live image of the passenger taken at the airport is compared to a gallery from CBP's TVS system.[45]

> *Workplace access control for employees and cleared guests (current use):* A company allows only employees and invited guests into their offices. The company employs security turnstiles in the building lobby that use FRT to allow entrance if the person matches a database of current employees and invited guests. In case of non-match, the person must visit the security desk.[46]

> *Workplace enforces visitor escort requirement (hypothetical):* A company requires that all visitors to its offices must be escorted by a staff member. Cameras deployed in the hallways can use FRT to look for any non-staff member who is not escorted by a staff member. If the system identifies an apparent violation, security guards rush to the location.

Although ID badges are used to identify non-cleared personnel who require escorts in classified security settings, the application of FRT as a mechanism for detection has not yet been deployed. Implementation of FRT for this case could be expanded to enforcing other escort requirements, as described in the use case above. FRT has also been used for access control of residential properties. This includes allowing the operation of elevators,[47] operating smart locks on the properties,[48] and ability to arm or disarm security systems.[49]

---

[42] Transportation Security Administration, n.d., "TSA PreCheck: Touchless Identity Solution," https://www.tsa.gov/biometrics-technology/evaluating-facial-identification-technology, accessed May 23, 2023.

[43] R. Santana and R. Gentilo, 2023, "TSA Is Testing Facial Recognition at More Airports," *AP News*, https://apnews.com/article/facial-recognition-airport-screening-tsa-d8b6397c02afe16602c8d34409d1451f.

[44] W. Chan, 2023, "TSA to Expand Facial Recognition Program to Over 400 Airports," *Fast Company*, https://www.fastcompany.com/90918235/tsa-facial-recognition-program-privacy.

[45] J. Doubleday, 2022, "CBP, TSA Expanding Facial Recognition for Traveler Identity Verification," *Federal News Network*, https://federalnewsnetwork.com/technology-main/2022/10/cbp-tsa-expanding-facial-recognition-for-traveler-identity-verification.

[46] R. Carriere, 2022, "Why Facial Recognition Makes Building Management Easier and Safter," *Facility Executive*, updated September 19, https://facilityexecutive.com/how-facial-recognition-makes-building-management-easier-and-safer.

[47] J.A. Kingson, 2023, "Elevators of the Future May Go Horizontal," *Axios*, https://www.axios.com/2023/01/04/artificial-intelligence-facial-recognition-elevators-otis-schindler-horizontal.

[48] S. Bajaj, 2023, "Best Face Recognition Door Locks," *Swiftlane*, updated June 20, https://swiftlane.com/blog/best-face-recognition-locks.

[49] Brinks Home, "Brinks Home 'Complete' Package," https://brinkshome.com/help-center/articles/360038959252-brinks-home-complete-package.

Automated school attendance (current use): A school places cameras in its classrooms and lecture halls and records students in the classroom. The school administration uses FRT to check the images of classroom attendance against a database of the student body in order to identify whether or not a student is attending class. The administration uses this attendance record to send truancy notices to parents of absentee students.[50,51,52]

Clocking in and out at work (current use): An employer deploys a camera near the employees' entrance. When a worker starts a shift, the camera scans their face and uses FRT to compare the captured image to a database of employee photos for record-keeping purposes.

Closely related to access control, the use of FRT by companies for employee time and attendance purposes is becoming very popular among employers.[53,54] Rather than clocking in and out of work using personal identity verification cards, personal identification numbers (PINs), and so on, this system simply uses a facial scan to identify the employee. An FRT system is contactless, reduces the need for replacement of lost cards, and would effectively eliminate instances of "buddy punching" and impersonation. In safety-critical industries, such as oil refining and chemical processing, FRT could be used in the event of an industrial accident to ensure that all individuals in the building have been accounted for in rescue efforts.[55]

Pharmacist access to controlled substance cabinet (current use): A nurse uses FRT to unlock a cabinet containing controlled medications. The image is compared to the employee's hospital ID. False negative (FN) results would prompt the nurse to

---

[50] Face-Six, n.d., "FA6 Class—Classroom Attendance. Face Recognition for School!" https://www.face-six.com/classroom-attendance-reinvented, accessed November 17, 2023.

[51] D. Samridhi and T. Patnaik, 2020, "Student Attendance System Using Face Recognition," *International Conference on Smart Electronics and Communication (ICOSEC)*, https://doi.org/10.1109/icosec49089.2020.9215441.

[52] A. Budiman, Fabian, R.A. Yaputera, S. Achmad, and A. Kurniawan, 2022, "Student Attendance with Face Recognition (LBPH or CNN): Systematic Literature Review," *Procedia Computer Science* 216:31–38, https://doi.org/10.1016/j.procs.2022.12.108.

[53] H. Kronk, 2021, "Facial Recognition Technology in the Workplace: Employers Use It, Workers Hate It, Regulation Is Coming for It," *Corporate Compliance Insights*, https://www.corporatecomplianceinsights.com/facial-recognition-technology-in-workplace.

[54] L. Rainie, M. Anderson, C. McClain, E.A. Vogels, and R. Gelles-Watnick, 2023, "Americans' Views on Use of Face Recognition in the Workplace," Pew Research Center, https://www.pewresearch.org/internet/2023/04/20/americans-views-on-use-of-face-recognition-in-the-workplace.

[55] R. Carriere, 2022, "Facial Recognition for Safer, More Efficient Factories," *Industry Today*, https://industrytoday.com/facial-recognition-for-safer-more-efficient-factories.

present an RFID (radio frequency identification) card to gain access to the locked cabinet and a review by hospital security.[56,57,58]

The use of FRT is becoming accepted for usage in health care facilities for a number of reasons. Facility security is a major driver for use of these systems not only for entrance into the facility but also to control access to sensitive parts of the facility. This includes access to pharmaceutical supplies and usage of critical equipment.

Identification to access public services (current use): An individual applying for government benefits has their face scanned, and FRT is used to confirm the individual's identity. Following authentication, the individual receives a photo ID denoting their receipt of benefits and their image is entered into a private database for later reference.

To address issues of fraud and inefficiency in public benefits access, many agencies have employed biometric-based systems for identity verification. Previously, fingerprints were obtained during the application process for nutrition assistance programs, such as the Supplemental Nutrition Assistance Program (SNAP) and Special Supplemental Nutrition Program for Women, Infants, and Children (WIC), as a method to deter fraud and prevent duplicate applications.[59] Most states and cities no longer require fingerprint verification due to the excessive cost and the increased stigmatization for applicants.[60] A recent report found that 22 labor agencies are using facial recognition for identity verification for unemployment insurance.[61] In this case, the applicants are asked to provide a government photo ID, as well as a video or "live selfie" that is compared using FRT.[62] Today, there is no evidence of biometrics, either facial recognition or fingerprinting, being used for identity verification to obtain other public services, such as SNAP, Temporary Aid to Needy Families Program, WIC, Medicaid, or Child Care Assistance Program.

---

[56] L. Pascu, 2020, "Cyberlink's FaceMe AI-Based Engine Integrated in AIoT iHospital Service Platform," *Biometric Update*, https://www.biometricupdate.com/202001/cyberlinks-faceme-ai-based-engine-integrated-in-aiot-ihospital-service-platform.

[57] L. Pascu, 2016, "MedixSafe Introduces Narcotic Safe with Facial Recognition," *Biometric Update*, https://www.biometricupdate.com/201603/medixsafe-introduces-narcotic-safe-with-facial-recognition.

[58] IDENTI Medical Data Sensing, 2023, "Secured Narcotics Cabinet: Medication Dispensing System for Controlled Substances and Narcotics," https://identimedical.com/narcotics-cabinet.

[59] C. Cournoyer, 2011, "Governments Abandon Fingerprinting for Food Stamps," *Governing*, https://www.governing.com/archive/governments-abandon-fingerprinting-food-stamp-recipients.html.

[60] C. Rodriguez, 2012, "The Clash Over Fingerprinting for Food Stamps," *NPR*, https://www.npr.org/2012/01/30/145905246/the-clash-over-fingerprinting-for-food-stamps.

[61] E.B. Sorrell, 2023, "Digital Authentication and Identity Proofing in Public Benefits Applications," updated November 19, https://www.digitalbenefitshub.org/publications/digital-authentication-and-identity-proofing-data.

[62] J. Buolamwimi, V. Ordóñez, J. Morgenstern, and E. Learned-Miller, 2020, *Facial Recognition Technologies: A Primer*, Cambridge, MA: Algorithmic Justice League, https://assets.website-files.com/5e027ca188c99e3515b404b7/5ed1002058516c11edc66a14_FRTsPrimerMay2020.pdf.

The Department of Housing and Urban Development (HUD) has used federal safety and security grants to help facilitate the purchase and installation of cameras equipped with FRT.[63] Although intended to prevent crime in public and HUD-assisted housing, video footage was used to identify, punish, and evict public housing residents, sometimes for minor violations of housing rules. It can also lead to the exclusion of unrecognized family members from the premises.

Check in for a flight (current use): An airline offers an opt-in feature allowing passengers to check in for their flight at an airport kiosk using FRT instead of showing ID or entering identification numbers. At the kiosk, the passenger pushes a button to trigger the FRT feature. If they are recognized, the kiosk greets them by name and initiates the check-in procedure. If not recognized, the kiosk asks for an identity document or flier number, then offers to opt the user in to future FRT, before continuing with check-in.[64]

In 2021, Delta was the first airline to introduce a digital identity program for TSA PreCheck members that offered "curb-to-gate" service at Detroit and then Atlanta airports.[65] The use of FRT in air travel has been extended to such functions as bag drop, security, and boarding.[66] With regard to COVID-19 pandemic protocols, FRTs have been deployed to improve social distancing procedures and increase the number of contactless interactions.

Face used to withdraw cash at ATM (current use outside the United States): A customer approaches an ATM and enters their PIN. The system asks the customer to face the camera so an image can be developed. Using the photo and PIN, the system verifies and validates the identity of the client and access is granted to the customer's accounts and cards. The customer can then withdraw cash or carry out other tasks available at the ATM. If a match is not made, the customer can present their ATM/debit card and use the ATM as normal.

Several banks, including CaixaBank (Spain), Shinhan Bank (South Korea), and Seven Bank (Japan), have rolled out facial recognition features for ATM withdrawals in

---

[63] D. MacMillan, 2023, "Eyes on the Poor: Cameras, Facial Recognition Watch Over Public Housing," *Washington Post*, https://www.washingtonpost.com/business/2023/05/16/surveillance-cameras-public-housing.
[64] *New York Times*, 2021, "Your Face Is, or Will Be, Your Boarding Pass," *New York Times*, https://www.nytimes.com/2021/12/07/travel/biometrics-airports-security.html.
[65] S. Writer, 2021, "Delta Launches First Domestic Digital Identity Test in U.S., Providing Touchless Curb-to-Gate Experience," *Delta News Hub*, https://news.delta.com/delta-launches-first-domestic-digital-identity-test-us-providing-touchless-curb-gate-experience; https://news.delta.com/deltas-exclusive-partnership-tsa-streamlines-check-security-atlanta.
[66] Newsdesk, 2021, "Delta Reveals First Dedicated TSA Precheck Lobby, Bag Drop," Travel Agent Central, https://news.delta.com/delta-reveals-first-ever-dedicated-tsa-precheckr-lobby-bag-drop.

recent years.[67,68] Although this technology has not yet been deployed by U.S. banks, the technology is readily available, and infrastructure is already in place to use facial recognition in lieu of PINs or, possibly, debit cards at ATMs. Facial recognition as a mechanism in lieu of passwords to access bank accounts on mobile devices has already been implemented as an opt-in convenience feature, utilizing the mobile device's facial recognition system. Hypothetically, this use case could be extended to include "self-check-out" purchases at a grocery store[69] or picking up prescriptions from a pharmacy.[70,71]

> *Amusement park season pass enforcement (current use outside the United States):* An amusement park sells annual passes. Pass holders have the option of entering the park through a special entrance, which uses FRT to check each entrant against a database of pass holders. In case of a non-match, the person is asked to use the public entrance where they will be asked to show ID.

Shanghai Disneyland[72] and Universal Studios in Singapore[73] have launched an opt-in facial recognition park entry app for its seasonal pass holders. Similar programs have been proposed at amusement parks in the United States but, as of the date of this report, none have been deployed.

PERSONAL DEVICE ACCESS

Facial recognition for use in security and access control for personal devices, such as smartphones, tablets, and laptop computers, has become increasingly common, allowing individuals to unlock their devices without having to type in their password. This ability is an opt-in feature, and the user can provide a non-biometric means for authentication and access.

---

[67] K. Flinders, 2020, "CaixaBank Introduces Facial Recognition ATMs," *Computer Weekly*, https://www.computerweekly.com/news/252484427/Caixabank-introduces-facial-recognition-ATMs.

[68] M. Borak, 2023, "South Korean Bank Rolls out ATM Withdrawals with Alchera Facial Recognition," *Biometric Update*, https://www.biometricupdate.com/202306/south-korean-bank-rolls-out-atm-withdrawals-with-alchera-facial-recognition.

[69] F. McFarland, 2023, "Huge Change Could Be Coming to Self-Checkout with Tech Used by Border Protection," *The U.S. Sun*, updated February 22, https://www.the-sun.com/money/7459309/shopping-facial-recognition-technology-border-protection.

[70] L. Biscaldi, ed., 2022, "Is Automated Prescription Pickup the Future of Pharmacy?" *Drug Topics* 166(3), https://www.drugtopics.com/view/is-automated-prescription-pickup-the-future-of-pharmacy-.

[71] J. Lee, 2017, "National Pharmacies Intros Facial Recognition at Australian Stores," *Biometric Update*, https://www.biometricupdate.com/201710/national-pharmacies-intros-facial-recognition-at-australian-stores.

[72] Shanghai Disney Resort, "Annual Pass Online Redemption and Facial Recognition Park Entry," https://www.shanghaidisneyresort.com/en/guest-services/facialrecognition, accessed November 17, 2023.

[73] Resorts World Sentosa Singapore, "Attractions Ticketing Terms and Conditions," https://www.rwsentosa.com/en/attractions/attractions-ticketing-terms-and-conditions, accessed November 17, 2023.

> *Unlock personal phone (current use):* A person opts in to using FRT to unlock their personal phone. Biometric references derived from images of their face are stored only in a secure area on the phone and will be deleted if the person later disables this feature.

An analogous scenario includes use of FRT to unlock and start a car, most notably implemented in the Genesis GV60.[74] Implementation of FRTs for use in personal device access is notable because it does not require interoperability—that is, both images are collected on the same camera and all components of the operation are specified and programmed by the developer. In this case, most attempts at recognition will be from the legitimate holder of the device, very few from an impostor. FNs will lead to a rejection and a prompt to retry. Consecutive FNs will often result in the phone prompting for authentication using an alternative modality (e.g., entry of a PIN or password); too many consecutive failures of the biometric and the alternative modality may result in the device being locked pending execution of an account recovery procedure that may need to be executed on a different device. A false positive would lead to unauthorized access to the phone, as FRT would incorrectly identify an individual as the legitimate device holder.

NONCONSENSUAL COMMERCIAL AND OTHER PRIVATE PURPOSES

> *Screening entrants to a venue based on a professional affiliation (current use):* The owner of a concert venue has legal conflict with a particular organization. They use FRT at all affiliated venue entrances to deny entrance to employees or legal representatives of this organization.[75]

Madison Square Garden and Radio City Music Hall have been reported to use FRT to identify lawyers who work at firms with pending litigation against them.[76] Lawyers have reported being escorted out of the venue, despite having purchased tickets and having never been involved in litigation against these venues. Although state and local regulations may vary, this is currently a legal practice in terms of federal law, as long as the venue does not discriminate against a class of explicitly protected citizens (e.g., age, race, gender, disability, religion, pregnancy, veteran status). Occupation and political affiliation, for instance, are not protected classes.

---

[74] C.J. Hubbard, 2022, "Genesis Launches Face Recognition for Cars," *CAR Magazine*, https://www.carmagazine.co.uk/car-news/tech/facial-recognition-key.
[75] K. Rhim, 2022, "Suing Madison Square Garden? Forget About Your Knicks Tickets," *New York Times*, https://www.nytimes.com/2022/10/13/sports/lawsuit-msg-lawyers-banned-knicks-rangers.html.
[76] K. Hill and C. Kilgannon, 2022, "Madison Square Garden Uses Facial Recognition to Ban Its Owner's Enemies," *New York Times*, updated January 3, https://www.nytimes.com/2022/12/22/nyregion/madison-square-garden-facial-recognition.html.

Personalized ads based on in-store browsing (hypothetical): A store places cameras at the entrance to capture the faces of entering customers. The store uses FRT to compare these images to stored customer profiles. When FRT identifies a match, the store uses its customer profile to generate personalized ads based on the customer's purchasing history.

Store identifies "high-value customers" (hypothetical): A luxury-goods store uses FRT at the store entrance to recognize "high-value customers" and dispatch a senior salesperson to assist each such customer.

Analogously, banks and other financial institutions outside of the United States have used FRTs to recognize their premium customers—identifying these customers upon entrance and providing them with premium services.[77] Facial recognition allows the banks to tailor their services specifically with their best customers in mind,[78,79] making a long-term association with the bank more likely. This use case can be easily extended to include car dealerships and upscale restaurants. Unless prohibited, it seems likely that it will only be a matter of time before stores will scan customers upon entry in order to personalize shopping experiences and marketing.

Individuals identified entering a health care facility (hypothetical): A third party uses a hidden camera and FRT to identify individuals entering a psychiatric clinic or a substance abuse treatment center, and the information is used to harass or blackmail individuals seeking treatment at the facility.

Concerns arise when surveillance FRT is used for private purposes to identify individuals who are present at a particular location. Although FRT systems are generally not available to individuals, there are services such as PimEyes that make it possible to identify individuals whose photos appear on the Internet. The resulting information could be sold to anyone, including private investigators, stalkers, foreign governments, or terrorist groups. Similar circumstances where unregulated facial recognition identification could be particularly problematic include, but are not limited to, attendees of a religious service at a synagogue, mosque, or church; protestors at a political rally; individuals under witness protection; or individuals seeking oncological, reproductive, or gender-affirming care.

---

[77] NEC, 2018, "NEC's Facial Recognition System Elevates Customer Experience at OCBC Bank," https://www.nec.com/en/press/201802/global_20180214_02.html#top.

[78] PYMNTS, 2023, "Mashreq Deploys Facial Recognition for Paperless Onboarding," updated July 6, https://www.pymnts.com/news/biometrics/2023/mashreq-deploys-electronic-facial-recognition-allow-paperless-onboarding.

[79] K. Flinders, 2023, "JP Morgan Pilots Palm and Face-Recognition Technology in U.S.," *ComputerWeekly*, March 27, https://www.computerweekly.com/news/365534158/JP-Morgan-pilots-palm-and-face-recognition-technology-in-US.

4

Equity, Privacy, Civil Liberties, Human Rights, and Governance

The implications of the use of facial recognition technology (FRT) for equity, privacy, civil liberties, and human rights are consequential, but the terms are contested, do not have fixed, universally accepted definitions, and overlap in important ways. In the following text, they are used to capture ways in which FRT can impact a core set of interests related to freedom from state and/or private surveillance, and hence control over personal information. Importantly, harm from surveillance is distinct from harms imposed by faulty or inadequate technical specifications and also distinct from harms that are measured in terms of their effects on diversity, equity, inclusion, and accessibility. In other words, although some potential FRT harms arise from errors or limitations in the technology, other potential harms arise and become more salient as the technology becomes more accurate and capable. Furthermore, it is important to emphasize that FRT can interfere with and substantially affect the values embodied in privacy, civil liberties, and human rights commitments without necessarily violating rights and obligations defined in current statutes or constitutional provisions.

This chapter considers the following related topics:

- The intersection of FRT with equity and race,
- Privacy and other civil liberties and human rights concerns associated with FRT use,
- Governance approaches for addressing these concerns,
- Governance issues raised by the use of FRT in criminal investigations, and
- Approaches for addressing wrongful FRT matches or overly intrusive deployment of FRT.

EQUITY, RACE, AND FACIAL RECOGNITION TECHNOLOGY

FRT intersects with equity and race in several key ways, as follows:

- *FRT manifests phenotypical variation in false positive (FP) match rates.* As discussed in Chapter 2, FRT developed in a particular region tends to over-represent particular phenotypes in its algorithmic training sets. Many FRT systems deployed in the United States are trained on imbalanced, disproportionately White, data sets. As a result, the systems yield consistently higher FP match rates when applied to racial minorities, including among populations that are Black, Native American, Asian American, and Pacific Islanders. Although overall error rates are, in absolute terms, very low in the best systems today under ideal conditions, individuals represented in these populations are nevertheless at higher risk of being erroneously identified by certain facial recognition systems.
- *FRT provides law enforcement with a powerful new tool for identifying individuals more rapidly, at a distance, and at greater scale and thus, depending on where and how it is used, has the potential to reinforce patterns or perceptions of elevated scrutiny by law enforcement and national security agencies, especially in marginalized communities.* Put bluntly, some communities may be more surveilled than others, and increased scrutiny can lead to neighborhoods being designated as high-crime areas, a feedback loop that can further justify use of FRT or other technologies that disproportionately affect marginalized communities. Moreover, the use of FRT has raised concerns in some communities—including Black, Hispanic, and Muslim communities—reflecting in part differential intensity of past interactions with law enforcement and other government authorities.
- *Several equity issues arise from the fact that reference galleries used by law enforcement—notably those based on mugshots—do not include every possible individual of interest for a scenario and may overrepresent and over-retain individuals from particular groups.* This means that
    - Differential intensity of policing can lead to differential frequency of law enforcement contacts, which leads to a differential rate of representation in law enforcement reference galleries. This effect is compounded by the fact that mugshots are not removed when cases are dropped or lead to acquittals.
    - Differential representation in galleries increases the probability of an FP match—that is, anyone in the gallery could become an FP match. Being

in the gallery at all is also a precondition for a false match based on lack of a high enough match score threshold. Conversely, not being in the gallery—because one has never had a law enforcement contact—not only makes the chance of an FP match zero but also makes the chance of a true match zero.

- *All six known cases where wrongful arrests have been made on the basis of FRT involve Black individuals identified using FRT.* These incidents likely represent a very small percentage of arrests involving FRT; comprehensive data on the prevalence of FRT use, how often FRT is implicated in arrests and convictions, or the total number of wrongful arrests that have occurred on the basis of FRT use do not exist. However, these cases have significance beyond what the numbers would suggest because they have occurred against a backdrop of deep and pervasive distrust by historically disadvantaged and other vulnerable populations of policing and because all of the reported wrongful arrests associated with the use of FRT have involved Black defendants. A brief summary of the cases follows:
  - Robert Williams was arrested in 2020 for a 2018 theft of watches on the basis of FRT identification made on the basis of a screen capture from security camera footage. He was detained for nearly 30 hours before being released on a personal bond. The detective working the case subsequently determined that Williams was not the person captured in the security camera footage.[1,2]
  - Nijeer Parks was arrested by police in New Jersey in 2019 after an erroneous FRT identification. He spent 11 days in jail after being charged with aggravated assault, unlawful weapons possession, using fake identification, shoplifting, marijuana, possession, resisting arrest, leaving the scene of a crime, and accusations of nearly striking a police officer with a car. He faced up to 25 years in jail, before he was able to produce evidence that he was 30 miles away when the crime occurred.[3]
  - Michael Oliver was arrested by Detroit police in 2019 on charges of stealing a cellphone. The investigator used FRT to identify Oliver as the suspect from video of the theft. It quickly became clear, however, that a

---

[1] T. Ryan-Mosley, 2021, "The New Lawsuit That Shows Facial Recognition Is Officially a Civil Rights Issue," *MIT Technology Review*, April 14, https://www.technologyreview.com/2021/04/14/1022676/robert-williams-facial-recognition-lawsuit-aclu-detroit-police.

[2] K. Johnson, 2022, "How Wrongful Arrests Based on AI Derailed 3 Men's Lives," *Wired*, March 7, https://www.wired.com/story/wrongful-arrests-ai-derailed-3-mens-lives.

[3] K. Hill, 2020, "Another Arrest, and Jail Time, Due to a Bad Facial Recognition Match," *New York Times*, December 29, https://www.nytimes.com/2020/12/29/technology/facial-recognition-misidentify-jail.html.

misidentification had occurred, because Oliver has visible tattoos on his arms while the individual filmed stealing the phone had none.[4]
- Randal Reid was arrested in 2022 driving to his mother's home in DeKalb County, Georgia, on a warrant issued in Louisiana on suspicion of using stolen credit cards. At the time of his arrest, Reid had never been to Louisiana. He was released after 6 days in detention.[5]
- Alonzo Sawyer was arrested in 2022 for allegedly assaulting a bus driver near Baltimore, Maryland, after FRT labeled him as a possible match to a suspect captured on closed-circuit television (CCTV) footage.[6]
- Porcha Woodruff was arrested and held for 11 hours in Detroit in 2023 for carjacking and robbery, despite the fact that she was 8 months pregnant at the time of the crime and the perpetrator was not.[7]

Perhaps the most detailed record has been developed by the press in the Williams case. The Williams arrest (see Box 4-1) and other cases illustrate that a combination of overconfidence in the technology, use of low-quality probe or gallery images, and poor institutional practices can lead to significant adverse impacts. In the six cases, the consequences have included false arrest and imprisonment, legal costs, interruption of normal activities of life and work, and loss of employment. Although six known wrongful arrests may seem like a small number, the lack of adequate data on law enforcement use of FRT makes it challenging to place these serious errors in a broader context. One cannot say with any confidence if these wrongful arrests are the only such examples, or if, instead, they are the tip of the iceberg. Nor can one say with assurance whether, or how much, the increased FP rate for phenotypically dark-skinned individuals contributed to these mistakes, although it is hard to accept that all six publicized wrongful arrests with FRT occurred with Black individuals as a matter of chance. In several of these cases, it appears that poor FRT procedures, inadequate training, and poor police investigative processes contributed to the erroneous arrests.

These intersections of FRT and race occur against a backdrop of historic and systemic racial biases that influence the development of technology. One commonly cited example with relevance to FRT is the history of film photography, which for many decades was calibrated for lighter skin tones (see Box 4-2). Although much work has

---

[4] E. Stokes, 2020, "Wrongful Arrest Exposes Racial Bias in Facial Recognition Technology." *CBS News*, November 19, https://www.cbsnews.com/news/detroit-facial-recognition-surveillance-camera-racial-bias-crime.

[5] K. Hill and R. Mac, 2023, "Thousands of Dollars for Something I Didn't Do," *New York Times*, March 31, https://www.nytimes.com/2023/03/31/technology/facial-recognition-false-arrests.html.

[6] K. Johnson, 2023, "Face Recognition Software Led to His Arrest. It Was Dead Wrong." *Wired*, February 28, https://www.wired.com/story/face-recognition-software-led-to-his-arrest-it-was-dead-wrong.

[7] K. Hill, 2023, "Eight Months Pregnant and Arrested After False Facial Recognition Match," *New York Times*, August 6, https://www.nytimes.com/2023/08/06/business/facial-recognition-false-arrest.html.

BOX 4-1 Robert Williams

On October 2, 2018, an unknown person's theft of several watches from a Shinola store in Detroit was captured on video surveillance cameras. A few days later, an analyst employed by a security firm contracted by Shinola provided image and video media related to the theft to the Detroit Police Department (DPD). Although the surveillance video footage was captured in high definition, the resulting frame grab was of very poor quality in terms of resolution and lighting. Additionally, the suspect wore a baseball cap that partially occluded key regions of the face. The images were ultimately sent to the Michigan State Police (MSP) to conduct a face recognition search.

MSP used a system developed by DataWorks, Inc., to search its repository of more than 49 million images, consisting of mugshots, driver's license photos, and state ID photos of Michigan residents. The system employs two face match engines, each returning a list of more than 200 face images of people with features most closely matched to the person of interest. The number of identical images appearing in both lists is unknown.

A driver's license photo of Mr. Williams surfaced in the ninth position in one of the candidate lists using one of the face matching algorithms. The license photo was from an expired license, not Mr. Williams's then-current license. The then-current license photo was also in the matching database but did not return as a candidate match. The second algorithm used in the search did not include any of Mr. Williams's license photos in its candidate list, nor did a search of a Federal Bureau of Investigation database.

The MSP image analyst selected the photo of Mr. Williams as a potential match, performed a morphological face comparison, and generated an investigative lead report containing the probe image and Mr. Williams's drivers' license photo, name, birthdate, and license number. DPD included a photo of Mr. Williams in a six-pack photo array and showed it to the representative of Shinola's security contractor to compare. The security company representative, who had never seen the suspect in person and had only watched the same security footage that was in DPD's possession, compared the photo of Mr. Williams to images from selected surveillance video frames and identified Mr. Williams as a match to the suspect.

Arrested on January 9, 2020, Mr. Williams was released on a personal bond after being detained for nearly 30 hours. Charges against Mr. Williams were subsequently dropped.

SOURCES: K. Hill, 2020, "Facial Recognition Tool Led to Black Man's Arrest. It Was Wrong," *New York Times*, June 25, p. A1, https://www.nytimes.com/2020/06/24/technology/facial-recognition-arrest.html; K. Hill, 2023, *Your Face Belongs to Us: A Secretive Startup's Quest to End Privacy as We Know It*, Random House, pp. 180–181; and E. Press, 2023, "Does A.I. Lead Police to Ignore Contradictory Evidence," *The New Yorker*, November 20, https://www.newyorker.com/magazine/2023/11/20/does-a-i-lead-police-to-ignore-contradictory-evidence.

> ## BOX 4-2 Race and Photography
>
> As discussed in Chapter 2, poor lighting or photography can reduce feature contrast. With less contrast, face detection algorithms may fail to detect a face in an image or, if a face is detected, the loss of facial detail can elevate facial recognition false negative match rates. This can be particularly problematic with images of faces of dark-skinned subjects.
>
> The underexposure of darker toned faces in photography has a troubling historical background. In the early days of color photography, film processing chemistry did not bring out certain red, yellow, and brown tones, because these tones were not seen as necessary where the market for photography was seen predominantly as light-skinned consumers. Skin tones were frequently calibrated using a stock image of a white woman. These test cards, known as "Shirley Cards" after the first name of the Kodak employee initially pictured, were widely used to calibrate skin tones in images produced on Kodak photographic printers. As a result, features of individuals with light skin were easily discernible in printed photographs, while features of individuals with dark skin were not.
>
> Efforts to correct this bias occurred in the 1970s and only because furniture and chocolate manufacturers complained that color film did not accurately render the colors of wood grain and chocolate. In the mid-1990s, as digital imaging went mainstream, Kodak responded by creating a multiracial Shirley Card with three women—one Black, one White, and one Asian. Since then, advancements in digital photography, such as better color balancing and image stabilization that reduces the need to use flash, have improved the presentation of darker skin tones, but even today, contrast will be worse when lighting is poor, and digital photography can still struggle with darker skin owing to biases in image processing algorithms, which themselves may reflect biases similar to those that afflicted film photography.
>
> SOURCES: C.M. Cook, J.J. Howard, Y.B. Sirotin, J.L. Tipton, and A.R. Vemury, 2019, "Demographic Effects in Facial Recognition and Their Dependence on Image Acquisition: An Evaluation of Eleven Commercial Systems," *IEEE Transactions on Biometrics, Behavior, and Identity Science* 1(1):32–41; NPR, 2014, "How Kodak's Shirley Cards Set Photography's Skin-Tone Standard," November 13, https://www.npr.org/2014/11/13/363517842/for-decades-kodak-s-shirley-cards-set-photography-s-skin-tone-standard; S. Lewis, 2019, "The Racial Bias Built into Photography," *New York Times*, April 25, https://www.nytimes.com/2019/04/25/lens/sarah-lewis-racial-bias-photography.html; and L. Roth, 2009, "Looking at Shirley, the Ultimate Norm: Colour Balance, Image Technologies, and Cognitive Equity," *Canadian Journal of Communication* 34(1):111–136, https://doi.org/10.22230/cjc.2009v34n1a2196.

been done in recent decades to address this bias, adequate lighting and contrast continue to be a challenge with darker skin tones.

CIVIL LIBERTIES, PRIVACY, HUMAN RIGHTS, AND FACIAL RECOGNITION TECHNOLOGY

"Civil liberties" is not a phrase found explicitly in the U.S. Constitution or any statute. It is used generally to capture a suite of fundamental rights and freedoms that protect individuals from unjust or oppressive government conduct. In the United States, civil liberties may be thought of as those rights associated with the federal and state constitutions. These include freedom of speech, freedom of assembly, freedom of the press, the right to privacy, and the right to due process when the government acts against a person. The term "human rights" is used globally to encompass a similar set of rights as captured in United Nations and other international agreements.

FRT has the potential to impact civil liberties and human rights because it changes the scale and cost of collecting detailed data about a person's movements and activities. Without FRT, a person can be momentarily observed in public, but it is expensive and difficult enough to make it practically impossible to track that person's movements extensively over time and space without a technical affordance that may be associated with an individual such as a cellphone or license plate. The proliferation of cameras can amplify the threat to civil liberties and privacy posed by FRT, including privately and law enforcement–operated CCTV cameras, doorbell cameras, and smartphones. Combined, these make it increasingly easy to identify people using images captured of their face. When FRT data are associated with space and time, the technology can become a means to evaluate a person's habits, patterns, and affiliations. Similar concerns have arisen with technologies such as license plate readers and cellphone location services. Some of the use cases identified in Chapter 3 may—depending on how they are implemented, used, and governed—implicate civil and human rights in concerning ways.

Privacy and Facial Recognition Technology

Privacy is commonly understood to include the right to control one's own personal information. This includes all forms of personal data, including, at least to some extent, personal movement, and behavior in the physical world and online. Of course, when people move around in public places, they can be observed. However, as was discussed in Chapter 1, FRT has the potential to further erode privacy in public spaces because it is inexpensive, scalable, and contactless and because it is very hard to avoid without masking one's face. Such identification and tracking impinge on privacy because of what it can reveal about a person's habits, behaviors, and affiliations that are reasonably not expected to be shared without permission. The potential to be tracked surreptitiously also unsettles widely shared expectations that one's movements will not be tracked or controlled in public spaces, at concert venues, at schools, etc., when one has not done anything unlawful. Defined in this way, privacy concerns itself less directly with the substance or subject matter of the information—whether about political or religious affiliations, financial data, medical information, sexual, or reproductive information—and more with the ability to preserve individual autonomy and freedom through the control of that information. This sense of autonomy, and hence control, includes the ability of persons to preserve their anonymity, as well as to control the circumstances and audiences to which personal information is revealed, at least to some extent. Importantly, most people understand that giving up a little control merely by moving through a public space does not mean that they have acquiesced to a complete loss of control. The fact that some inferences can be drawn about a person who moves in public does not mean that there are no privacy interests to defend.

Privacy guarantees can be found in federal constitutional provisions related to freedom of speech and association, protection against unreasonable search and seizure, and substantive due process rights protecting privacy, family, and intimate associations. State constitutions can also provide privacy protections, sometimes to a greater degree than the U.S. Constitution. Federal and state statutes, such as the Privacy Act[8] and the Health Insurance Portability and Accountability Act,[9] can also provide legal protections of privacy interests against both the civil and government actors.

Indiscriminate use of FRT in public and quasi-public places can have significant impacts for privacy and related civil liberties. Indeed, the collection of images in public places that could be subject to FRT may deter people from exercising their civil rights. FRT can be used to scan lawful protests or other large gatherings for potential or known threats. However, in the process, data would be collected on individuals who raise no legitimate law enforcement concerns. The use of FRT to identify individuals in other public or quasi-public spaces raises similar concerns—especially absent regulation or other controls on how such information is collected, stored, and used. These concerns may be heightened in locations associated with religious, political, or medical practice. Moreover, the use of FRT in public or quasi-public spaces might also have particularly adverse consequences for the privacy of individuals such as informants, undercover agents, protected witnesses, and victims of abuse. Furthermore, collected data could conceivably be sold to foreign actors, increasing exposure for U.S. citizens while abroad.

---

[8] Privacy Act of 1974, as amended, 5 U.S.C. § 552a.
[9] Health Insurance Portability and Accountability Act. Pub. L. No. 104-191, § 264, 110 Stat.1936.

Individuals can also apply FRT to an image using an online service such as PimEyes, allowing them to identify individuals in images obtained on the Internet or captured using smartphone, doorbell, and other cameras. Widespread availability of such a capability alters expectations about anonymity in public and private places and is especially troubling because it can be used to identify individuals for harassment, intimidation, stalking, or other abuse. Already, one can take a photograph of someone standing nearby or across the street, run it through PimEyes, and receive a small gallery of likely matches, permitting the potential identification of one stranger by another.

Privacy concerns have also been raised regarding how the data used in FRT systems are gathered. Although many law enforcement agencies likely rely on galleries of mugshots or driver's license photos, the leading private FRT vendor, Clearview AI, compiles its FRT gallery by collecting public images from the Internet, including social media, without consent from the platform or the individuals pictured. To date, Clearview AI has built a database of more than 30 billion images. This practice has met with pushback from some governments. In 2022, the United Kingdom's privacy watchdog, the Information Commissioner's Office, ordered the company to "delete all data belonging to UK residents," becoming the fourth country—following Australia, France, and Italy—to do so. Following a lawsuit filed by the American Civil Liberties Union under Illinois's Biometric Information Privacy Act, which creates a private right of action, Clearview AI signed a settlement that permanently barred the company from selling its database to most private businesses. Despite this opposition, the company's FRT systems are still frequently used by law enforcement across the country. According to Clearview AI, as of 2021, the company counts 3,100 law enforcement agencies as customers, along with the Army and the Air Force. In March 2023, the company reported that its database has been used nearly 1 million times by U.S. law enforcement.

In addition to general privacy concerns raised by inclusion in large databases, data in such centralized repositories are highly sensitive and may be an attractive target for exfiltration by third parties, including criminals and foreign governments. Indeed, it is potentially highly useful to adversaries of the United States.[10] Protecting the security of such data is essential to protecting the national security of the United States and the privacy and civil liberties of Americans.

---

[10] For example, a 2015 breach of data held by the Office of Personnel Management resulted in the exposure of data impacting 22.1 million Americans. The federal government and its data contractor agreed to a $63 million settlement with individuals whose personally identifiable information was stolen. See E. Katz, 2022, "A Judge Has Finalized the $63M OPM Hack Settlement. Feds Now Have Two Months to Sign Up for Damages," *Government Executive*, October 26, https://www.govexec.com/pay-benefits/2022/10/judge-finalized-63m-opm-hack-settlement-feds-two-months-damages/378950.

Other Civil Liberties Concerns

FRT has been used by business owners to monitor customers and identify potential shoplifters, resulting in several cases of businesses using a false match from an FRT system as the basis for excluding or removing an individual. The prospect of authorities and property owners detaining an individual, or denying access to a store, venue, or other establishment solely on the basis of an FRT match, without recourse, may in many circumstances be viewed as an unwanted expansion of state or private powers. For example, a 2020 investigation from Reuters found that Rite Aid had deployed FRT systems at more than 60 stores in predominantly low-income minority neighborhoods to assist in loss prevention.[11] The investigation further identified cases in which false matches generated by the FRT system resulted in an individual being wrongfully asked to leave the store on suspicion of shoplifting by Rite Aid management.

Human Rights and International Perspectives

Human rights are rights enjoyed by all persons. The Universal Declaration of Human Rights,[12] a key document setting forth fundamental human rights worthy of universal protection, was adopted in 1948 by the United Nations General Assembly. It provides a basic framework for later conventions and other legal instruments that have emerged in the development of international human rights law. They include the right to be free from "arbitrary interference with [one's] privacy, family, home, or correspondence, [and from] attacks upon [one's] honour and reputation." Human rights principles are expected to be respected by both government and private actors. The United Nations' Guiding Principles on Business and Human Rights, for instance, state that "business enterprises should respect human rights."

The use of FRT is being questioned beyond the United States. In 2018, Big Brother Watch, a civil society organization, investigated the use of FRT by police departments in the United Kingdom, demonstrating how FRT "disproportionately misidentif[ies] minorities and women, and 95 percent of [UK] police's matches have misidentified individuals."[13] In 2019, researchers with the University of Essex Human Rights Centre published a report on the deployment of live facial recognition (LFR) technology by the London Metropolitan Police Service, noting a "lack of publicly available guidance on the use of LFR."[14] In 2022, Chatham House published a report documenting a swift increase

---

[11] J. Dastin, 2020, "Special Report: Rite Aid Deployed Facial Recognition Systems," *Reuters*, July 28, https://www.reuters.com/article/us-usa-riteaid-software-specialreport-idUSKCN24T1HL.

[12] United Nations General Assembly, 1948, "The Universal Declaration of Human Rights," https://www.un.org/en/about-us/universal-declaration-of-human-rights.

[13] Big Brother Watch, 2018, *Face Off: The Lawless Growth of Facial Recognition in UK Policing*, London, England: Big Brother Watch, https://bigbrotherwatch.org.uk/wp-content/uploads/2018/05/Face-Off-final-digital-1.pdf.

[14] P. Fussey and D. Murray, 2019, *Independent Report on the London Metropolitan Police Service's Trial of Live Facial Recognition Technology*, Colchester, England: University of Essex Human Rights Centre.

in "the deployment of facial recognition in public spaces for police surveillance" in Latin America without adequate regulations.[15] In China, where the deployment of FRT has been particularly extensive (e.g., to track Uighurs through their daily lives in Xinjiang province), the Supreme People's Court, in a "joint stance with Beijing's top government bodies," called for stronger consumer privacy protections from "unwarranted face tracking," introducing new guidelines in 2021 requiring commercial venues to obtain "consent from consumers to use facial recognition," to limit FRT use to "what is necessary," and to protect consumer's data.[16,17]

THE GOVERNANCE OF FACIAL RECOGNITION TECHNOLOGY

The impacts of FRT on equity, privacy, and civil rights are greatest when images are indiscriminately collected, stored, and analyzed with little or no input, regulation, or oversight from individuals, communities, civil society organizations, or governmental bodies. FRT raises difficult questions for governance because it raises many novel and complex legal questions. The complexity arises from the following:

- Many actors are involved in FRT system design and development, the collection of images for training template extraction models, and deployment and use of FRT capabilities. Some of these activities raise unsettled legal questions that depend, in part, on where and how FRT is used (e.g., in a public or commercial space, by a private or a government actor, etc.); and
- Regulation of FRT might take place at different levels of government (i.e., national, state, and local). Furthermore, at any given level, FRT might be subject to regulation by existing general laws (e.g., related to intellectual property, privacy, law enforcement), technology specific law or regulation, or both.

There are several pathways for federal regulatory action on FRT. First, a court might interpret the U.S. Constitution as providing limits on the government's use of FRT or as providing constraints on state or national authority to regulate FRT. Constitutional law on both questions is unsettled, and there are no directly applicable or dispositive

---

[15] C. Caeiro, 2022, *Regulating Facial Recognition in Latin America: Policy Lessons from Police Surveillance in Buenos Aires and São Paulo*, London, England: Royal Institute of International Affairs, https://doi.org/10.55317/9781784135409.

[16] E. Dou, 2021, "China Built the World's Largest Facial Recognition System. Now, It's Getting Camera-Shy," *Washington Post*, July 30, https://www.washingtonpost.com/world/facial-recognition-china-tech-data/2021/07/30/404c2e96-f049-11eb-81b2-9b7061a582d8_story.html.

[17] National Institute of Standards and Technology (NIST), 2020, "Facial Recognition Technology (FRT)," February 6, https://www.nist.gov/speech-testimony/facial-recognition-technology-frt-0.

Supreme Court rulings. See the discussion below on how constitutional protections might apply.

Second, Congress could enact a statute directly regulating FRT. However, although there are legislative proposals to regulate FRT, no legislation on the regulation of FRT has been enacted into law.

Third, a federal agency could issue a regulation or initiate an enforcement action under a statute of general application (i.e., not related to FRT) to address both state and private uses of FRT. Alternatively, guidelines for use by federal agencies could be developed, potentially as directed by an executive order.

Legislative Approaches to the Governance of Facial Recognition Technology

U.S. Federal Law

Currently, no federal statute or regulation imposes a general constraint on the public or private use of FRT. However, there are existing agency authorities or legislative mandates that may have applicability to FRT in specific instances.

The Federal Trade Commission (FTC), for example, has used its authority under Section 5 of the Federal Trade Commission Act to regulate "unfair or deceptive acts or practices in or affecting commerce" to take action against a photo-app developer that allegedly deceived consumers about its use of FRT[18]—and could potentially address other FRT-related acts or practices.

Federal laws requiring privacy impact assessments and system of record notices impose transparency requirements on federal agencies that use FRT. For example, a May 2016 Government Accountability Office (GAO) report identified privacy and transparency concerns with the Federal Bureau of Investigation's (FBI's) use of FRT. In response, the FBI expedited work on system of record notices (which notify the public about the existence of systems and the types of data they collect) and privacy impact assessments (which examine how systems collect, store, manage, and share personal information).[19]

Another avenue for federal action is the establishment of rules for the procurement and funding of FRT, and non-binding standard-setting activities (such as those of the National Institute of Standards and Technology [NIST][20]).

---

[18] Federal Trade Commission, 2021, "FTC Finalizes Settlement with Photo App Developer Related to Misuse of Facial Recognition Technology," September 18, https://www.ftc.gov/news-events/news/press-releases/2021/05/ftc-finalizes-settlement-photo-app-developer-related-misuse-facial-recognition-technology.

[19] Government Accountability Office, 2019, "Face Recognition Technology: DOJ and FBI Have Taken Some Actions in Response to GAO Recommendations to Ensure Privacy and Accuracy, But Additional Work Remains," https://www.gao.gov/products/gao-19-579t.

[20] NIST, 2020, "Facial Recognition Technology (FRT)," February 6, https://www.nist.gov/speech-testimony/facial-recognition-technology-frt-0.

Training data used by FRT algorithms may be protected by contract or privacy law, but the scope of these protections is unclear. Social media platforms alleged that Clearview AI violated its terms of service by collecting facial images from the Internet. In response, Clearview AI asserted a First Amendment right to collect the images.[21] Some have asserted that U.S. copyright law protects against the collection and use of facial images from the Internet. Open questions include whether such activities fall under the fair use exception or special provisions that apply to providers of search engines and similar tools.

Under the Supreme Court's interpretation of the free speech clause of the First Amendment, regulation of the commercial collection and use of data may be prohibited. In *Sorrell v. IMS Health*, the Court held that the sale, disclosure, and use of pharmacy records was First Amendment speech.[22] The idea that information is speech[23] gives private actors powerful support for the assertion that FRT development and deployment cannot be regulated. Nevertheless, the First Amendment only applies to private speech. *Sorrell*, therefore, does not preclude federal regulation of state actors such as state and municipal police agencies. It may, however, be constitutionally impossible to regulate elements of the private market—including firms that market their services aggressively to police.

A few bills have focused on regulating FRT, illustrating concerns of some members of Congress. One example of proposed, FRT-specific legislation is the Facial Recognition Act of 2022 (H.R. 9061), which was introduced but never received a committee vote.[24] The bill focuses on the use of FRT by law enforcement, eschewing a categorical ban on FRT, and instead would require, among other constraints, a judge-authorized warrant before conducting facial recognition searches, notice to individuals subject to FRT searches, and a ban on FRT searches using databases of illegally obtained photographs. The bill would also require law enforcement agencies to annually submit data about their use of FRT for audit by the GAO and require that FRT systems be tested annually using NIST's benchmark for facial recognition for law enforcement. The bill also includes provisions for redress—including suppression of FRT results and any derivative evidence—in the event of improper use of FRT. Another example of legislation that has been introduced but thus far not acted on is a series of similar bills calling for a moratorium on federal law enforcement use of FRT, introduced most recently as the Facial Recognition and Biometric Technology Moratorium Act of 2023 (S. 681). More generally, proposed legislation to regulate artificial intelligence (AI) would address issues such as bias and civil rights compliance, and would, if enacted, also have implications for the regulation of FRT.

---

[21] See B.E. Devany, 2022, "Clearview AI's First Amendment: A Dangerous Reality?" *Texas Law Review* 101(2):473–507.
[22] *Sorrell v. IMS Health Inc.*, 564 U.S. 552, 570 (2011).
[23] Ibid.
[24] See https://www.congress.gov/bill/117th-congress/house-bill/9061/text?s=1&r=20, H.R. 9061.

U.S. State Regulations

Illinois was the first state to regulate FRT through the 2008 Biometric Information Privacy Act (BIPA).[25] BIPA regulates "the collection, use, safeguarding, handling, storage, retention, and destruction of biometric identifiers and information." It prohibits private parties from collecting biometric identifiers or using information derived from biometric identifiers to create individual profiles without notification, consent, and specified disclosures. Furthermore, BIPA prohibits the sale of collected biometric identifiers and requires private parties to make public their data retention and destruction policies. Similar laws were enacted in Arkansas, California, Texas, and Washington.

Illinois's and California's statutes allow for a private right of action, but the costs of civil litigation mean that it is often not feasible for individuals to bring suit. Only aggregate litigation, such as class action lawsuits, will have positive expected value before litigation begins. As a result, in the absence of nonprofit legal assistance, individual remedies under these statutes will likely rarely be pursued.

Lawsuits have been filed under BIPA but have ended in settlements rather than judgments. In 2020, for instance, Facebook settled a lawsuit alleging that its creation of face profiles violated Illinois's biometric privacy law, changing its use of FRTs as a result.[26] In 2021, Clearview AI settled a lawsuit under the same state law.[27] Clearview AI suggested that it has a First Amendment defense to BIPA liability,[28] but the soundness of this argument is unsettled because the case was not adjudicated.

Another potential avenue for state legislation is to regulate the use of FRT by law enforcement. For example, Maryland Senate Bill 192 would limit the use of FRT to serious crimes and threats to public safety or national security and prohibit use of FRT as the sole basis to establish probable cause.

U.S. Municipal Regulations

Local regulations governing surveillance technologies, including FRT, can mandate public approval of the acquisition and use of these technologies, require transparency or prohibit non-disclosure agreements, and confer legal standing to citizens to challenge violations of these rules. They can also impose notice requirements on private companies that use FRT as part of their business. Most municipalities have not, however, taken action to regulate the use of FRT. In the instances where they have, efforts have typically taken two forms: (1) the creation of administrative agencies with responsibility for public

---

[25] 740 Ill. Comp. Stat. Ann. 14/15(b).

[26] N. Singer and M. Isaac, 2020, "Facebook to Pay $550 Million to Settle Facial Recognition Suit," *New York Times*, January 29, https://www.nytimes.com/2020/01/29/technology/facebook-privacy-lawsuit-earnings.html.

[27] R. Mac and K. Hill, 2022, "Clearview AI Settles Suit and Agrees to Limit Sales of Facial Recognition Database," *New York Times*, May 9, https://www.nytimes.com/2022/05/09/technology/clearview-ai-suit.html.

[28] B.E. Devany, 2022, "Clearview AI's First Amendment: A Dangerous Reality?" *Texas Law Review* 101(2):473–507.

surveillance technologies, review of annual reports on the use of these technologies, and new regulations and (2) city councils with legislative and administrative functions that establish procedures for the acquisition and use of surveillance technologies.

There have been moves by municipal jurisdictions to categorically ban FRT. In 2019, the city of San Francisco banned the use of FRT by its public agencies.[29] Under the city's administrative code, it is unlawful for any public agency to "obtain, retain, access, or use" any FRT on "city-issued software or a city-issued product or device" or any information obtained from FRT.[30]

Since 2016, the American Civil Liberties Union (ACLU) has been active in promoting a model bill for local governments interested in regulating surveillance technology in public hands. Called the Community Control Over Police Surveillance (CCOPS), the model bill requires city council approval before the "acquiring or borrowing [of] new surveillance technology" and the issuance of a "Surveillance Impact and Surveillance Use Policy" for any proposed technology. In 2023, at least 22 local governments, including Boston, Massachusetts; New York, New York; Detroit, Michigan; San Francisco, California; and San Diego, California, have adopted surveillance technology regulations using the ACLU model as a template but local governments have made significant alterations from the model bill.[31]

The city of Oakland, California, has often been cited as a model for local governance of surveillance technologies.[32] Oakland enacted surveillance technology regulations and created a separate Privacy Advisory Commission to advise the Oakland City Council on privacy issues. The council and the commission share responsibility for approving new purchases or uses of surveillance technologies by public agencies. If, for instance, the Oakland Police Department seeks to adopt or change a use policy for a surveillance technology, it must notify the commission and then present a surveillance impact policy and use report. The commission conducts public hearings, creates reports, and makes recommendations to the city council regarding the city's acquisition and use of technology that "collects, stores, transmits, handles or processes citizen data."[33] The city council has final decision-making authority.[34]

---

[29] Oakland, California, and Somerville, Massachusetts, have also passed local ordinances banning the use of FRT by public agencies. See Fight for the Future, "Ban Facial Recognition Map," https://www.banfacialrecognition.com/map, accessed November 17, 2023.

[30] City and County of San Francisco, 2019, "Administrative Code–Acquisition of Surveillance Technology: Board of Supervisors Approval of Surveillance Technology Policy," Section 19B.2(d), https://sfgov.legistar.com/View.ashx?M=F&ID=7206781&GUID=38D37061-4D87-4A94-9AB3-CB113656159A.

[31] M. Fidler, 2020, "Local Police Surveillance and the Administrative Fourth Amendment," Santa Clara, Computer and High Technology Law Journal, Aug. 2, p. 546, http://dx.doi.org/10.2139/ssrn.3201113.

[32] Ibid.

[33] "Privacy Advisory Commission," City of Oakland, California, https://www.oaklandca.gov/boards-commissions/privacy-advisory-board, accessed November 17, 2023.

[34] City of Oakland, "Chapter 9.64: Regulations on City's Acquisition and Use of Surveillance Technology."

Regulation of FRT at the local government level faces some obstacles. Local ordinances will result in policies and regulations that vary from city to city. In addition, the creation of a specialized administrative agency or regulations for surveillance technologies like FRT requires resources, including access to technical expertise, that many municipalities do not have access to. Outside of large cities, an approach that emphasizes municipalities will necessarily leave oversight gaps.

There has also been some pushback against regulations limiting or banning use of FRT by law enforcement, citing concerns about crime.

International Law

Other countries are also grappling with whether and how to govern FRT. Perhaps the most ambitious attempt at regulation is contained within the European Artificial Intelligence Act, which would complement the General Data Protection Regulation and the Law Enforcement Directive of the European Union. The European Parliament adopted its negotiating position on the act in June 2023, and a provisional agreement of the European Parliament and the European Council on the final form of the act was announced on December 9, 2023—but the final text of the act had not been released as of this writing. The press release states the following regarding biometric identification systems:

> Negotiators agreed on a series of safeguards and narrow exceptions for the use of biometric identification systems (RBI) in publicly accessible spaces for law enforcement purposes, subject to prior judicial authorisation and for strictly defined lists of crime. "Post-remote" RBI would be used strictly in the targeted search of a person convicted or suspected of having committed a serious crime.[35]

Constitutional Protections

Several provisions of the U.S. Constitution have potential relevance to FRT. The application of these provisions to the use of FRT is currently being studied, contested, and litigated. The most important are the Fourth Amendment's protection against unreasonable searches and seizures, the Fifth Amendment due process right, the equality components of the Fifth and the Fourteenth Amendments, and the First Amendment's free speech clause.[36] It is important to note, however, that almost every constitutional prohibition applies only to "state action."[37] Although it can sometimes be unclear where state action ends and private action begins, the Constitution generally only applies when

---

[35] European Parliament, 2023, "Artificial Intelligence Act: Deal on Comprehensive Rules for Trustworthy AI," press release, December 9, https://www.europarl.europa.eu/news/en/press-room/20231206IPR15699/artificial-intelligence-act-deal-on-comprehensive-rules-for-trustworthy-ai.

[36] It is also possible to imagine religious liberty challenges to the mandatory use of face verification. These, however, would not constitute a general regulation of the technology, and so are not addressed here.

[37] *Manhattan Cmty. Access Corp. v. Halleck*, 139 S. Ct. 1921, 1928 (2019).

there is a federal or state official directly acting (and not, say, when a private actor voluntarily supplies information [such as footage or an identification] to a government actor). Nor do these protections apply when private actors act toward other private actors in a manner that would be unconstitutional had the government acted in the same manner.

Fourth Amendment

The Fourth Amendment is commonly associated with privacy from state intrusion, especially from law enforcement. It would be appropriate to presume that the amendment speaks to state use of FRT, but this presumption may not hold. The courts have not ruled on the question of whether the state's collection of facial images is a Fourth Amendment "search." Unless this threshold condition is met, the amendment would not apply.

In the context of the Fourth Amendment, a search is understood to have occurred when there is a violation of a "person's reasonable expectation of privacy."[38] It is not clear, however, whether a person has a reasonable expectation of identification privacy in a public setting. If something is already "in plain view, neither its observation nor its seizure would involve any invasion of privacy."[39] This plain view exception reflects the intuition that, when something can be lawfully observed by an official, there is no reasonable expectation of privacy. This suggests that one does not have a reasonable expectation of privacy in one's facial features when in public, but how should one think about this set of issues when one's facial features can be used to make an identification or to track movement?

The question thus arises as to whether the state collection of facial data in a public setting could *ever* trigger Fourth Amendment scrutiny. Given the reasonable expectations test and the plain view exception, it is doubtful that federal courts would proscribe the general use of public surveillance cameras[40] (although there is disagreement among lower federal courts as to whether long-term surveillance of a home using a pole-mounted surveillance camera constitutes a Fourth Amendment search).[41] But even this carve-out would have a limited effect on FRT use because it would apply only to a small fraction of public video footage. Because the Fourth Amendment turns on how information is collected by the state actor—rather than how it is used—analysis of public surveillance footage for either identification purposes (with or without FRT) likely does not raise Fourth Amendment concerns.[42]

---

[38] *Katz v. United States*, 380 U.S. 347, 360 (1967) (Harlan, J. concurring).

[39] *Horton v. California*, 496 U.S. 128, 133 (1990).

[40] *Leaders of a Beautiful Struggle v. Baltimore Police Dep't*, 979 F.3d 219, 231 (4th Cir. 2020), on reh'g en banc, 2 F.4th 330 (4th Cir. 2021) ("Precedent suggests law enforcement can use security cameras without violating the Fourth Amendment.").

[41] See, for example, *United States v. Tuggle*, 4 F.4th 505, 511 (7th Cir. 2021), cert. denied, 212 L. Ed. 2d 7, 142 S. Ct. 1107 (2022).

[42] The Fourth Amendment applies only to government "searches" and "seizures." The surveillance of a public place is neither. When the government does not acquire information directly from a suspect, but from a third party, the Fourth Amendment is typically not implicated. For an exception, discussed later, see *Carpenter v. United States*, 138 S. Ct. 2206 (2018).

Importantly, the Fourth Amendment does not provide protection from all warrantless searches by the state. The Supreme Court has carved out, for example, an exception to the warrant requirement of the Fourth Amendment at the border. At the border, officials "have more than merely an investigative law enforcement role,"[43] and greater power to search. Federal courts have hence upheld suspicionless searches of cellphones and laptop searches at the border that would be illegal if conducted during the course of ordinary policing.[44] Furthermore, the Supreme Court has developed an "administrative search" doctrine that permits searches without probable cause or warrants for many regulatory purposes.[45] Due to such carve-outs, the Fourth Amendment often provides weak privacy protection outside a crime-control context, and its application to immigration enforcement, in particular, is highly context dependent.

All this suggests that the Fourth Amendment might offer only limited protections against the use of FRT, particularly when deployed in proximity to the border. Importantly, the government could also employ FRT far from the border (e.g., around workplaces likely employing noncitizens or on a subway to aid in the search for undocumented persons).

Another important question relates to instances where the government relies on a private party to deploy FRT. Under the "third-party" doctrine, Fourth Amendment protections do not apply when the government acquires records about a person from a third party—such as a bank or a telephone company.[46] Hence, the state's use of a private security firm's footage would not trigger a Fourth Amendment concern because the state did not obtain data directly from the suspect. This distinction, however, is not absolute. In 2018, the U.S. Supreme Court created an exception to the third-party doctrine for the use of cellphone location data to pinpoint a suspect's physical whereabouts over time.

In *Carpenter v. United States*, the Supreme Court ruled that "individuals have a reasonable expectation of privacy in the whole of their physical movements."[47] Reasoning from the Framers' ambition to "place obstacles in the way of a too permeating police surveillance," it expressed particular concern over the risk of "near perfect surveillance" by which police could—retroactively, if need be—"retrace a person's whereabouts."[48] The Court also emphasized the "deeply revealing nature" of location data—its "depth, breadth, and comprehensive reach" and "the inescapable and automatic nature of its collection."[49] The Court frankly grappled with the way in which the diffusion of new

---

[43] *United States v. Montoya de Hernandez*, 473 U.S. 531, 544 (1985).
[44] *Alasaad v. Mayorkas*, 988 F.3d 8, 19 (1st Cir.), cert. denied sub nom. *Merch. v. Mayorkas*, 141 S. Ct. 2858 (2021) (holding, along with several circuit courts, that "basic border searches are routine searches and need not be supported by reasonable suspicion").
[45] See, for example, *New York v. Burger*, 482 U.S. 691, 703 (1987) (exempting regulatory inspections of automobile dismantling businesses from warrant and probable cause requirements).
[46] *United States v. Miller*, 425 U.S. 435, 443 (1976).
[47] *Carpenter v. United States*, 138 S. Ct. 2206, 201 L. Ed. 2d 507 (2018).
[48] Ibid.
[49] Ibid.

surveillance tools, coupled to novel analytic strategies, can expand the state's power to acquire personal information. Many of these concerns, though, can be extended easily to private actors that can also tap into broad information-gathering powers, even if the Fourth Amendment does not apply.

To date, *Carpenter* has not been extended to the use of FRT. Nevertheless, the U.S. Court of Appeals for the Ninth Circuit invoked *Carpenter* to reason that "the development of a face template using facial-recognition technology without consent ... invades an individual's private affairs and concrete interests."[50] Legal scholars have developed broad readings of *Carpenter* that would lead to more extensive regulation of FRT.[51]

The most natural application of *Carpenter* would be to FRT-based surveillance tools that focus on "prolonged tracking that can reveal intimate details through habits and patterns."[52] This understanding of *Carpenter* might mean that some facial identification use cases would be subject to a warrant requirement under the Fourth Amendment; it is less likely to include facial verification use cases.

A separate question arises as to whether an FRT "match" made by law enforcement may by itself constitute sufficient cause for either a brief investigative detention or an arrest. The Fourth Amendment requires that arrests must be based on probable cause and that "no warrants shall issue, but upon probable cause, supported by oath or affirmation," a legal standard that the U.S. Supreme Court has described as a "practical, nontechnical conception."[53] Law enforcement may subject a person to a brief investigatory detention based on the less demanding Fourth Amendment standard of reasonable suspicion.[54] "Terry stops," for example, allow police to detain a person briefly based on a reasonable suspicion of involvement in criminal activity, and arrests also permit the police to engage in other investigative activities, including searches incident to an arrest.

Equality Under the Fifth and Fourteenth Amendments

A persistent concern about FRT relates to potential differential effects on racial and ethnic groups. The Fifth and the Fourteenth Amendments prohibit certain actions taken on the basis of race by the federal government and the states, respectively. Constitutional equality law, however, is not triggered by the creation of racial or ethnic disparities.[55] A violation instead requires a particular showing of intent. A government decision-maker must have "selected or reaffirmed a particular course of action at least in part

---

[50] *Patel v. Facebook, Inc.*, 932 F.3d 1264, 1273 (9th Cir. 2019) (finding standing on this basis).
[51] Even the most ambitious of these accounts recognizes "constitutional gaps in protective coverage requiring legislative action." A.G. Ferguson, 2019, "Facial Recognition and the Fourth Amendment," *105 Minnesota Law Review 1105*, October 21, https://doi.org/10.2139/ssrn.3473423.
[52] *Leaders of a Beautiful Struggle v. Baltimore Police Dep't*, 2 F.4th 330, 341 (4th Cir. 2021).
[53] *Illinois v. Gates*, 462 U.S. 213, 231 (1983).
[54] *Terry v. Ohio*, 392 U.S. 1 (1968).
[55] *Washington v. Davis*, 426 U.S. 229, 240 (1976).

'because of,' not merely 'in spite of,' its adverse effects upon an identifiable group."[56] Especially in criminal and immigration cases, the Court has created a set of presumptions and procedural rules that make it exceedingly hard for most litigants to prove improper intent.[57] In addition, the Supreme Court has carved out a near-categorical prohibition on official decision-makers taking explicit account of race in their decision-making protocols.[58] This is commonly known as the colorblindness mandate. In effect, these rules prohibit a narrow class of intentional or explicitly race-conscious or racially directed actions. Where the government uses a criterion (e.g., residential zip code) that closely correlates with racial identity, its application is less clear.

Under current constitutional equality doctrine, FRT is unlikely to face successful challenges. Specific FRT instruments may have racially disparate effects, but this is typically not because of an intention to harm a minority. Nor is race used in an explicit criterion in matching. Constitutional equality law, moreover, would not be violated if a policing agency were to use an FRT with racial disparate effects unless its choice were demonstrated to be "because of" and not merely "in spite of" these disparities. It would be very difficult under current law for a plaintiff to satisfy this burden. Moreover, it is possible that certain race-conscious measures to mitigate those disparities may run into constitutional objections.[59] For example, the Court has invalidated an official decision to reject the outcome of an employment test with racially disparate effects. It reasoned that abandoning an action because of racial disparities itself was a problematic race-conscious action.[60] It would seem, therefore, that an agency concerned at avoiding racial disparities as a consequence of an FRT instrument would be advised to act up front by purchasing a tool that did not evince those gaps, rather than by trying to rectify such disparities after the fact.

Internal Law Enforcement Guidelines

Local law enforcement agencies can set administrative governance principles by establishing departmental rules and guidelines on the use of FRT. As an illustration of local law enforcement guidelines, consider the New York Police Department's (NYPD's) internal guidelines for FRT use.[61] FRT may only be used by NYPD for a specified set of authorized uses, including to identify a person when there is "a basis to believe that such individual has committed, is committing, or is about to commit a crime."[62] The NYPD

---

[56] *Personnel Admr. v. Feeney*, 442 U.S. 256, 279 (1979).
[57] A.Z. Huq, 2019, "What Is Discriminatory Intent?" *Cornell Law Review* 103(5), https://scholarship.law.cornell.edu/clr/vol103/iss5/4.
[58] *Parents Involved in Cmty. Sch. v. Seattle Sch. Dist. No. 1*, 551 U.S. 701 (2007).
[59] A.Z. Huq, 2018, "Racial Equity in Algorithmic Criminal Justice," *Duke Law Journal* 68(663).
[60] *Ricci v. DeStefano*, 557 U.S. 557 (2009); see also A.Z. Huq, 2019, "Racial Equity in Algorithmic Criminal Justice," *Duke Law Journal* 68(1043) (discussing applications to other criminal justice algorithms).
[61] NYPD Patrol Guide, Procedure No. 212-129 (3/1/2020).
[62] Ibid.

guidelines also state that the determination of a possible FRT match alone "does not constitute probable cause to effect an arrest, or obtain an arrest or search warrant."[63]

Local law enforcement–developed rules and guidelines, unlike administrative governance by city councils or local agencies, may be vulnerable to questions of legitimacy and independence. Moreover, there is a risk that some interests will not be systematically represented within existing review and decision-making processes, including the interests of communities most intensively subject to FRT tools. These concerns can be mitigated at least in part by deliberate efforts to engage with stakeholders in their development.

Federal law enforcement agencies have also created internal agency guidance on the use of FRT, which are subject to review by agency and department general counsels and leadership. For example, the FBI's Facial Analysis, Comparison, and Evaluation Services Unit (which uses reference databases containing hundreds of millions of faces, including driver's license photos from more than a dozen states) and the Next Generation Identification-Interstate Photo System (which processes thousands of requests from state and local law enforcement agencies per month)[64] are subject to internal regulations (though these are not public).

Governance by Private Entities

When local governments fail or choose not to adopt policies and regulations on FRT use, technology vendors can become the default rulemaking bodies. Vendors may impose non-disclosure agreements with contracting municipalities, and thus create problems of transparency and accountability. Contract terms imposed by vendors might, for instance, specify that data generated by FRT belong to the vendor and not the public agency or the city.[65] This raises important transparency and equity concerns.

Private technology vendors may decide, as a policy matter, not to incorporate FRT into tools offered to law enforcement agencies. In 2019, Axon, the country's largest supplier of police body-worn cameras and software, announced that it would impose a moratorium on FRT use in its devices.[66] However, such self-regulation has limits. The decision creates no legally enforceable rights or remedies for individuals or third parties in the event that the company violates its own policies. Axon could reverse course at any time. Furthermore, law enforcement agencies using Axon products could transfer data collected from the company's products to a third party for FRT analysis.

---

[63] Ibid.

[64] Congressional Research Service, 2020, "Federal Law Enforcement Use of Facial Recognition Technology," CRS R46586, https://sgp.fas.org/crs/misc/R46586.pdf.

[65] S. Gordon, "Milwaukee Committed to Shotspotter But Outcomes, Data Remain Elusive," Wisconsin Public Radio, January 20, https://www.wpr.org/milwaukee-committed-shotspotter-outcomes-data-remain-elusive. (Reporting that data generated by gunshot detection system ShotSpotter are owned by the company and that "SST's ownership of the data is written into the contracts it signs with law enforcement.")

[66] C. Warzel, 2019, "A Major Police Body Cam Company Just Banned Facial Recognition," *New York Times*, June 27, https://www.nytimes.com/2019/06/27/opinion/police-cam-facial-recognition.html.

FACIAL RECOGNITION TECHNOLOGY IN CRIMINAL INVESTIGATIONS AND TRIALS

Once an FRT match has been made, there are a number of scenarios where the resulting match could be invoked or applied (e.g., in a criminal investigation or in the course of the proceedings of a criminal trial).

In criminal investigations, current best practice is to use FRT as one component of investigative leads. This practice is reflected, for example, in guidelines from the Facial Identification Scientific Working Group (FISWG), whose members include a number of federal, state, and local law enforcement agencies as well as law enforcement agencies in Europe, the Americas, and Australia. The introduction to FISWG's document on minimum training criteria for personnel who conduct facial comparisons using FRT states:

> An automated FRS typically provides a list of candidates from a database in response to a facial image query. The user of an FRS and the personnel reviewing the results are required to be aware of the major elements and limitations of the facial comparison discipline and training in the use of available tools. Results from an automated FRS are used as investigative leads only and should be used in conjunction with additional resources.[67]

In legal settings, the use of the results of an FRT match is also subject to strict procedural constraints.

Fifth Amendment

Pursuant to the Fifth Amendment, prosecutors in a criminal action have a due process obligation to disclose to a defendant all evidence that is "favorable" and "material either to guilt or to punishment."[68] If a prosecution were to rely on evidence from an FRT match, the Fifth Amendment may require the prosecution to disclose "evidence of police misuse of facial recognition and poor algorithm quality."[69] At least one state appeals court has determined that the government was obligated to disclose detailed information about the FRT tool used to identify a suspect. Especially because FRT is a "novel and untested technology," the court ordered the disclosure of the "identity, design, specifications, and operation of the program or programs used for analysis."[70]

---

[67] Facial Identification Scientific Work Group, 2021, "Minimum Training Criteria When Using Facial Recognition Systems," Version 1.0, October 22, https://fiswg.org/fiswg_min_training_criteria_when_using_fr_systems_v1.0_2021.10.22.pdf.
[68] *Brady v. Maryland*, 373 U.S. 83, 87 (1963).
[69] J. Brown, 2022, "We Don't All Look the Same: Police Use of Facial Recognition and the Brady Rule," *Federal Communications Law Journal* 74(3):329–346.
[70] *State v. Arteaga*, 476 N.J. Super 36, *61 (App. Div. 2023).

Evidentiary Issues

Although many of the currently known instances of FRT use involve the development of investigative leads, courts will need to determine whether and how FRT matches may be admitted as evidence. To resolve a disputed issue about novel scientific or technical information, a court may permit a party to introduce testimony by an expert witness. In assessing the reliability of expert testimony, a court may consider a variety of factors, including amenability to testing, whether there is a known error rate and standards governing the use of the technique in practice, whether the technique has been subject to peer review in scientific publications or otherwise, and whether the technique or method has general acceptance in the relevant scientific community.[71] At least one state court has observed that there is "no agreement in a relevant community of technological experts that [FRT] matches are sufficiently reliable to be used in court as identification evidence,"[72] but given the general willingness to permit prosecutors to introduce expert evidence in court, it is likely that at some point, courts may determine that FRT is sufficiently valid and reliable to be introduced as evidence of identification. It is also possible that the fact that FRT has played a role in an investigation may be permitted, not as independent evidence of identification, but as part of the "res gestae"—the background circumstance and explanatory narrative describing the events that led to the arrest.

If the result of an FRT were to be introduced in court as evidence of identification, it would be critical for the court to determine both that the technology itself is adequately valid and reliable—that it has, as the President's Committee of Advisors on Science and Technology report on forensic science put it, "foundational validity"—and that it was applied reliably by an appropriately trained, competent analyst in this particular instance.[73] Determining validity may also raise issues of access to technical details about the surveillance instrument, which may in turn raise access issues given potential non-disclosure agreements or trade secrets.[74]

---

[71] *Daubert v. Merrell Dow Pharmaceuticals*, 509 U.S. 579 (1993); *Kumho Tire Co. v. Carmichael*, 526 U.S. 137 (1999).

[72] *People v. Reyes*, 133 N.Y.S.3d 433, 436-437 (N.Y. County 2020).

[73] Executive Office of the President, President's Council of Advisors on Science and Technology, 2016, *Forensic Science in Criminal Courts: Ensuring Scientific Validity of Feature-Comparison Methods*, Report to the President, Washington, DC, https://obamawhitehouse.archives.gov/sites/default/files/microsites/ostp/PCAST/pcast_forensic_science_report_final.pdf.

[74] R. Wexler, 2017, "Life, Liberty, and Trade Secrets: Intellectual Property in the Criminal Justice System," *Stanford Law Review*.

ADDRESSING WRONGFUL MATCHES AND INTRUSIVE DEPLOYMENT OF FACIAL RECOGNITION TECHNOLOGY

Increasing use of FRTs in the public and private sectors raises questions about legal and administrative remedies for harms caused by the use of FRT. Courts may be asked to consider whether some FRT uses give rise to civil liability under traditional causes of action, and legislatures may wish to consider whether new legislation providing causes of action is warranted.

Individuals may seek legal relief in cases of mistaken FRT matches. Those harmed by mistaken FRT matches may rely on existing federal or civil rights causes of action, although their exact applicability in this context of FRT is not well defined. Federal law offers damages remedies and the possibility of injunctive relief when a constitutional rule such as the Fourth Amendment is violated. Furthermore, criminal defendants can ask that evidence gathered in violation of the Fourth Amendment be suppressed. But such remedies are, in practice, often not available because of a complex network of rules that limit the availability of damages or suppression except in instances where a government official has committed a particularly obvious and egregious violation of constitutional law. With new technologies, persons asserting a constitutional right must often point to previous judicial rulings to show specifically that a constitutional violation was especially egregious; it is not enough to point to a general, foundational ruling. But the hurdles to relief mean such rulings are sparse on the ground.[75]

As a practical matter, state statutes currently offer the only meaningful relief for individuals harmed by FRT. While, as noted above, federal agencies such as the FTC might offer remedies for deceptive commercial practices and violations of federal statutory law, the remedies are often designed to prevent future illegal behavior, not to make whole those harmed by a new technology.[76]

When FRT use by private actors is perceived as unduly invasive, individuals may seek remedies in the form of common law–based privacy torts against those actors. Most states recognize tort causes of action—for example, for the public disclosure of private facts, intrusion upon seclusion, false light (spreading falsehoods about an individual), and appropriation of name or likeness. For instance, a person who experiences what is perceived as nonconsensual and highly invasive use of FRT by private actors may rely on the privacy tort of "intrusion upon seclusion." Although there is no widely recognized general expectation of privacy in public, some courts have suggested there may be limited exceptions in ways that might apply to the FRT context. For instance, the New York

---

[75] A.Z. Huq, 2015, "Judicial Independence and the Rationing of Constitutional Remedies," *Duke Law Journal* 65(1).
[76] The Everalbum settlement is an example of that sort of remedy.

State Court of Appeals opined that "overzealous" surveillance may be actionable when the information sought is "of a confidential nature" and the defendant's conduct was "unreasonably intrusive."[77]

For policymakers and organizations seeking to deploy and use facial recognition appropriately and safely, public transparency about the circumstances under which FRT is used is important. Furthermore, the disclosure of information regarding the technical performance of the deployed FRT system can create pressure on organizations to use top-performing algorithms and foster public confidence in the accuracy of these systems. Clear guidance on factors to consider when deploying FRT can help organizations identify use cases that may require more stringent safeguards. Training and certification programs for the personnel using and reviewing system outputs can ensure a uniform baseline competence.

Systems can be designed to strengthen privacy protections, particularly with regard to the storage of reference galleries and probe images. For instance, reference galleries should always store templates, which are derived from face images, rather than the images themselves. Meanwhile, to prevent inappropriate use of probe images for searches beyond pre-defined operational needs, systems can be configured to automatically delete captured probe images at the end of a set, publicly disclosed retention period.

Policy measures can help alleviate concerns related to the use of FRT. For instance, robust notice and consent practices could be enacted to notify individuals when their images might be stored in a reference gallery or used for training purposes and would give meaningful potential to opt out of image collection. Furthermore, deploying organizations and developers could develop data policies that limit data collection to absolutely necessary purposes, strictly govern how those data are to be used, and limit the long-term retention and sharing of facial image data. In crafting policy, policymakers might consider the context in which FRT is deployed. For instance, policymakers could ask whether a given deployment results in a greater scope, scale, and persistence of record-keeping than existed without the use of FRT. Measures might be taken to ensure that there is adequate justification for a given deployment of FRT, that consideration is given to who will bear responsibility for protecting privacy, and that privacy protections for certain vulnerable groups are appropriate (e.g., domestic violence survivors, individuals enrolled in witness protection, and other groups who may be endangered by the sharing of their whereabouts). Privacy impact assessments are used by the federal government and other organizations as a structured approach for considering such questions and making the analysis available to the public.

---

[77] *Nader v. General Motors Corp.*, 255 N.E. 2d. 560, 567 (Ct. App. N.Y. 1970).

Several mitigation measures might help address civil and human rights concerns. For instance, disclosure requirements could be enacted wherein those deploying FRT must clearly and publicly state that FRT is in use and for what purposes. Industry codes of conduct could be developed to promote best practices. Tools such as export controls might be employed to restrict access of FRT to authoritarian regimes.

5

Conclusions and Recommendations

Facial recognition technology (FRT) has matured into a powerful technology for identification and identity verification. Some uses offer convenience, efficiency, or enhanced safety, while others—including ones already deployed in the United States—are troubling and raise significant equity, privacy, and civil liberties concerns that have not been resolved by U.S. courts or legislatures.

Concerns about the use of FRT arise from two (non-exclusive) factors that require different analysis and merit different policy responses:

- *Concerns about poor performance of the technology*—for example, unacceptable false positive (FP) or false negative (FN) rates or unacceptable variation of these rates across demographic groups.
- *Concerns about problematic use or misuse of the technology*—for example, technology with acceptable performance sometimes produces societally undesirable outcomes as a result of inadequate procedures or training for operating, evaluating, or making decisions using FRT or because FRT is deliberately used to achieve an outcome not foreseen by developers or vendors.

That is, some concerns about FRT can be addressed by improving the technology while others require changes to procedures or training, restrictions on when or how FRT is used, or regulation of the conduct that FRT enables. Furthermore, some uses of FRT may well cause such concern that they should be not only regulated but prohibited.

TECHNICAL PERFORMANCE AND STANDARDS

Current top-performing facial recognition algorithms provide prompt, high-confidence matches when the probe image is obtained cooperatively and when the reference image is of high quality. Under these conditions and using today's best face recognition algorithms, 99.9 percent of searches with a sufficiently clear face image will return the correct matching entry in a government database of 12 million identities in under a second.

Two key performance metrics are FP and FN match rates.

- An FP occurs when the technology erroneously associates the template of a probe image with a template in the gallery. In some cases, the individual photographed in the probe image may not even have a corresponding template in the reference gallery. Recent stories of false arrests enabled by FRT typically involve an FP match, as the image of an innocent person in the gallery is incorrectly matched to a probe image of a suspected perpetrator. As the size of reference galleries or the rate of queries increases, the possibility of a false match grows, because there are more potential templates that can return a high similarity score to a probe face. The FP rate will be very high for twins and other individuals with a close familial resemblance to the probe face.
- An FN occurs when a probe image of an individual whose image is contained in the reference gallery returns no matches. For instance, when a passenger on a departing airplane is asked to present their face for recognition at the boarding gate, an FN may occur when the technology erroneously fails to identify the passenger in the gallery of individuals on the flight manifest. In this case, an FN may require the traveler to show photo identification.

Matching performance will be worse when the probe image is obtained under suboptimal conditions (e.g., poor lighting) or when the reference image is outdated or of low resolution or contrast. Nevertheless, with the best available algorithms, as long as both the eyes in a face can be automatically detected, a probe image can be matched to an individual with more than 99 percent accuracy.[1] In many cases, even if only one eye can be detected, an image of an individual can still be matched with high accuracy; even profile-view images can often be correctly matched.

Much progress has been made in recent years to characterize, understand, and mitigate phenotypical disparities in the accuracy of FRT results. However, these

---

[1] See the latest NIST FRTE report on 1:N matching, https://pages.nist.gov/frvt/reports/1N/frvt_1N_report.pdf.

performance differentials have not been entirely eliminated, even in the most accurate existing algorithms. FRT still performs less well for individuals with certain phenotypes, including those typically distinguished on the basis of race, ethnicity, or gender.

Tests show that FN rate differentials are extremely small if both the probe and reference images are of high quality, but the differentials can become significant if they are not. FN matches occur when the similarity score between two different images of one person is low. Causes include changes in appearance and loss of detail from poor image contrast. FN match rates vary across algorithms and have been measured to be higher by as much as a factor of 3 in women, Africans, and African Americans than in Whites. The most accurate algorithms also generally have the lowest demographic variance. FN match rate disparities are highest in applications where the photographic conditions cannot be controlled and can be reduced with better photography and better comparison algorithms. The consequences of an FN match include a failure to identify the subject of an investigation or the need for an individual to identify themselves in another way, such as by presenting identity documents. Rate disparities mean, for example, that the burden of presenting identification falls disproportionately on some groups of individuals—including groups that have been historically disadvantaged and marginalized. Although this additional time and inconvenience may be seemingly small in a single instance, the aggregate impacts to individuals who repeatedly encounter it and to groups disproportionately affected can be large.

FP matches occur when the similarity score between images of two different people is high. (The likelihood of an FP can thus be reduced with a higher similarity threshold.) Higher FP match rates are seen with women, older subjects, and—for FRT algorithms designed and trained in the West—individuals of East Asian, South Asian, and African descent. However, some Chinese-developed algorithms have the lowest FP rates for East Asian subjects. FP match rate differences occur even when the images are of very high quality and can vary across demographic groups markedly and contrary to the intent of the developer. FP match rate disparities can be reduced using more diverse data to train models used to create templates from facial images or model training with a loss function that more evenly clusters but separates demographic groups. The applications most affected by FP match rate differentials are those using large galleries and where most searches are for individuals who are not present in the gallery. FP rate disparities will mean that members of some groups bear an unequal burden of, for example, being falsely identified as the target of an investigation.

Tests also show that for identity verification (one-to-one comparison) algorithms, the FP match rates for certain demographic groups, when using even the best performing facial recognition algorithms designed in Western countries and trained mostly on White faces, are relatively higher (albeit very low in absolute terms), even if both the probe and reference images are of high quality.

A final concern with FPs is that as the size of reference galleries or the rate of queries increases, the possibility of an FP match grows, as there are more potential templates that can return a high similarity score to a probe face. Some face recognition algorithms, however, adjust similarity scores in an attempt to make the FP match rate independent of the gallery size.

> **RECOMMENDATION 1: The federal government should take prompt action along the lines of Recommendations 1-1 through 1-6 to mitigate against potential harms of facial recognition technology and lay the groundwork for more comprehensive action.**

> **RECOMMENDATION 1-1: The National Institute of Standards and Technology should sustain a vigorous program of facial recognition technology testing and evaluation to drive continued improvements in accuracy and reduction in demographic biases.**

Testing and standards are a valuable tool for driving performance improvements and establishing appropriate testing protocols and performance benchmarks, providing a firmer basis for justified public confidence, for example, by establishing an agreed-on baseline of performance that a technology must meet before it is deployed. The National Institute of Standards and Technology's (NIST's) Facial Recognition Technology Evaluation has proven to be a valuable tool for assessing and thereby propelling advances in FRT performance, including by increasing accuracy and reducing demographic differentials. This work, and the trust it has engendered, provide the foundation for NIST to take on an expanded role in developing needed standards in such areas as evaluating and reporting on performance, minimum image quality, data security, and quality control.

> **RECOMMENDATION 1-2: The federal government, together with national and international standards organizations (or an industry consortium with robust government oversight), should establish**
> a. **Industry-wide standards for evaluating and reporting on the performance—including accuracy and demographic variation—of facial recognition technology products for private or public use.**
> b. **A tiered set of profiles that define the minimum quality for probe and reference images, acceptable overall false positive and false negative rates, and acceptable thresholds for accuracy variation across different phenotypes for applications of different sensitivity levels. It would be up to those creating guidance, standards, or**

regulations to select the appropriate profile for the application in question.

c. **Methods for evaluating false positive match rates for probe images captured by closed-circuit television or other low-resolution cameras (which have been implicated in erroneous arrests of several Black individuals).**

d. **Process standards in such areas as data security and quality control.**

NIST would be a logical home for such activities within the federal government given its role in measurement and standards generally and FRT evaluation specifically.

RISK MANAGEMENT FRAMEWORK

Organizations deploying FRTs face a complex set of trade-offs and considerations as they seek to use the technology fairly and effectively. To help manage these complex trade-offs around privacy, equity, civil liberties, and technical performance, a framework that is specified in advance can help users identify and manage risks, define appropriate measures to protect privacy, ensure transparency and effective human oversight, and identify and mitigate concerns around equity. A framework can similarly assist bodies charged with oversight of FRTs, whether governmental agencies or civil society organizations, in making decisions about where the use of FRTs is appropriate and where it should be constrained. Such a framework could also form the basis for future mandatory disclosure laws or regulations.

> **RECOMMENDATION 1-5: The federal government should establish a program to develop and refine a risk management framework to help organizations identify and mitigate the risks of proposed facial recognition technology applications with regard to performance, equity, privacy, civil liberties, and effective governance.**

Risk management frameworks are a valuable tool for identifying and managing risks, defining appropriate measures to protect privacy, ensuring transparency and effective human oversight, and identifying and mitigating concerns around equity. A risk management framework could also form the basis for future mandatory disclosure laws or regulations.[2] Current examples of federally defined risk management frameworks

---

[2] A recent Federal Trade Commission statement calls for assessment of risks.

include NIST's Cybersecurity Framework and NIST's Artificial Intelligence Risk Management Framework. NIST would be a logical organization to be charged with developing this framework given its prominent role in FRT testing and evaluation as well as in developing risk management frameworks for other technologies.

A framework for the use of FRT might address the following:

1. Technical performance
    1.1 Does the FRT perform with the accuracy of current state-of-the-art systems? Does it perform with adequate accuracy for the intended application?
    1.2 Does the FRT have differential accuracy rates across different demographic groups of concern that are as low as current state-of-the-art systems? Is the differential adequately low for the intended application?
    1.3 Does it conform to the prevailing technical standards at the time of deployment, such as those specified by NIST?
    1.4 Do the subject and reference images conform to appropriate standards for image quality to support a match at the intended level of confidence?
    1.5 Does the FRT system adequately communicate to users the confidence of a reported match?
    1.6 Does it offer users with sufficient context information to mitigate other kinds of error?
2. Equity, privacy, and civil liberties
    2.1 Equity
        2.1.1 Does use of the FRT system result in statistically and materially significantly different treatment for different demographic groups? Is this attributable to technical characteristics (1.2) or other factors?
        2.1.2 What steps have been taken to mitigate equity risks associated with using the technology in a specific use case?
        2.1.3 How are any of these differences assessed, reported, and disclosed?
        2.1.4 What training is being conducted to ensure that when in use, users understand FRT impacts on federally protected groups?
        2.1.5 What is the pre-assessment in FRT's design for risk mitigation around equity concerns?
        2.1.6 Who makes up the training data and what are the contexts in which the data are collected (e.g., public or private databases)?
        2.1.7 Who is participating in the model's design and evaluating outcomes for equity?

- 2.1.8 Are the data extracted representative to avert potential errors in positive identification?
- 2.1.9 What documentation is being gathered to audit for civil rights compliance and equity?
- 2.1.10 What are the apparent and unintended sociotechnical outcomes of the FRT?

2.2 Privacy
- 2.2.1 Privacy of faces used in training the template extraction model
    - 2.2.1.1 Are privacy-preserving methods used, and if not what other measures are taken to protect the privacy of people whose images were used?
    - 2.2.1.2 Are data used for training the template extraction model acquired with consent and in compliance with relevant user agreements? Will the data used for this be purchased or sold without consent of individuals in the data set?
    - 2.2.1.3 Was the database constructed with data obtained in compliance with the terms of service for the data source?
- 2.2.2 Are best practices for data security and integrity of FRT training data and reference databases—including adequately protecting information in FRT training data sets and reference databases from exfiltration and misuse—being followed?
- 2.2.3 Have appropriate data collection, disclosure, use, and retention policies for both subject and reference images and templates been put in place to limit, for example, inappropriate use of probe images for searches beyond pre-defined operational needs?
- 2.2.4 Does the use of FRT significantly increase the scope or scale of the identification being performed?
    - 2.2.4.1 In a world before FRT, would you have been identified in this setting?
    - 2.2.4.2 Does the use of FRT allow for identification on a scale that would have been impractical without FRT?
    - 2.2.4.3 Is the reference database being searched appropriate to the application? Is the search being performed in the smallest possible closed group?
    - 2.2.4.4 Would there have been a record kept of the identification, and for how long? Is this record-keeping consistent with the record-keeping without FRT?

- 2.2.4.5 If FRT is being used for forensic purposes, is the record kept consistent with current forensic practice?
- 2.2.5 Does the use of FRT lead to any other adverse privacy impacts?

2.3 Civil liberties
- 2.3.1 Is the outcome of this FRT being used to control access to a public benefit or service, and if so, does it accord with due process norms?
- 2.3.2 Would the deployment of FRT in a given use case have a reasonably foreseeable negative impact on the exercise of civil rights, such as free speech or assembly, whether by individuals or groups?
- 2.3.3 Is the use of FRT in compliance with existing civil rights laws?

2.4 Surveillance (which implicates equity, privacy, and civil liberties concerns)
- 2.4.1 Is the FRT being used by government actors, commercial interests, or private individuals? (Government and commercial uses of FRT may be more amenable to regulation and oversight than use by private individuals.)
- 2.4.2 Is FRT applied to images collected retrospectively, live, or prospectively? (The use of retrospective images may mean that the subjects' images were collected without notice or consent that FRT use was contemplated at the time of collection.)
- 2.4.3 Is FRT applied for mass surveillance or individually targeted use? Is its use limited or indefinite in duration? (Indiscriminate or indefinite use of FRT on large crowds may pose greater threats to civil rights and civil liberties than the use of FRT to identify one or several individuals based on individualized suspicion.)
- 2.4.4 Is the FRT application susceptible to uses constituting harassment, abuse, or new opportunities for criminal or civil harm? (Current or future FRT applications may, for instance, invite private individuals to identify persons in sensitive situations, track their movements, or endanger their safety.)
- 2.4.5 Is the FRT application intended to be used covertly or transparently, particularly in places traditionally deemed public? If notice is provided, is the context such that it is reasonable to expect people to be able to make a choice about using such locations?
- 2.4.6 Is the FRT application being used for exclusionary, adversarial, or punitive purposes, or is it likely to be so used?

- 2.4.7 Is the FRT application being used against communities or in places that have historically experienced abusive or disproportionate surveillance practices, or is it likely to be so used?
- 2.4.8 Do those who believe they have been subjected to a mistaken FRT match have a means of redress (e.g., administrative complaints, legal causes of action, etc.)?

3. Governance
    - 3.1 Public interest or legitimate business purpose
        - 3.1.1 For government uses, is there an important public interest? Does FRT clearly enable that interest to be better served? What costs are imposed, and has every effort been made to minimize them?
        - 3.1.2 For commercial and other private uses, is there a legitimate business purpose?
        - 3.1.3 Is FRT being used for cases beyond the stated purpose?
        - 3.1.4 What safeguards exist against unauthorized uses?
    - 3.2 Decision-making about deployment
        - 3.2.1 Who decides whether and how to deploy the technology?
        - 3.2.2 Who will be operating the technology?
        - 3.2.3 Does the organization deploying and operating the FRT bear the risks, or are the risks externalized?
    - 3.3 Community and stakeholder engagement
        - 3.3.1 What consultation is done with the public at large or specific potentially affected groups?
        - 3.2.2 Has the consultation engaged with a sufficiently large and representative set of individuals?
        - 3.2.3 Have the results of the consultation been meaningfully considered (and at a minimum, have any changes been made) in determining whether deployment is appropriate, and whether safeguards are needed?
    - 3.4 Safeguards and oversight
        - 3.4.1 Who is responsible for ensuring that appropriate safeguards are in place and being followed?
        - 3.4.2 Does the system produce a record that can be used ex post for system verification and evaluation?
        - 3.4.3 Are safeguards, such as access controls or audit trails, in place to prevent unintended use—and if such use occurs, to impose appropriate penalties?

- 3.4.4 Does the system keep biometric data separate from non-biometric data?
- 3.4.5 Does the entity using FRT adhere to quality management and assurance practices per the ISO 9000 standards?

3.5 Disclosure

- 3.5.1 Is there meaningful public disclosure about where, when, and for what purpose the system is used, or has a clear and compelling justification been offered for why such disclosure is not needed?
- 3.5.2 Is there a clear and publicly accessible data retention policy for both subject and reference images? Will the data be sold or transferred to another entity? Is this narrowly tailored to the stated purpose, and is this properly disclosed?
- 3.5.3 In data retention systems, are sufficient guardrails established regarding the sharing and retention of images for purposes other than the reason for the original retention?

3.6 Consent

- 3.6.1 Is the FRT system opt-in? If it is opt-in, is the opt-in mechanism uncoerced? If it is an opt-out application, is the opt-out mechanism meaningful? (Analogous questions arise with both consent for the use of an FRT system and consent for one's face to be included in a reference gallery.)
- 3.6.2 If FRT is mandatory (i.e., there is no opt-in or opt-out), is there a clear and compelling justification?
- 3.6.3 Are individuals in practice able to consent to the proposed use? Are individuals reasonably able to understand the implications of consent? If individuals were given the option not to consent, what fraction of them would refuse in this application?
- 3.6.4 Are there procedures in place for persons who cannot consent by law (e.g., minors, etc.)?
- 3.6.5 Were the reference images captured appropriately—that is, with consent or per legitimate government authority? Is there a protocol for eliminating reference images that are gathered without proper and lawful authority?

3.7 Training

- 3.7.1 What sort of capabilities or competencies does the operator of an FRT system need to demonstrate? How are these updated as new capabilities are added to an FRT system?

- 3.7.2 Do the training or certification regimes adequately mitigate the risks of the system usage?
- 3.8 Human-in-the-loop
  - 3.8.1 Is an identified individual responsible for all significant decisions or actions made on the basis of an FRT match result?
- 3.9 Accountability
  - 3.9.1 Which is the expected/positive outcome or adverse outcome for an individual? What is the cost or consequence to an individual of an adverse outcome?
  - 3.9.2 Are appropriate (i.e., commensurate with cost/consequence) recourse/redress mechanisms available to individuals who will experience adverse outcomes?
  - 3.9.3 Does the organization using FRT have a mechanism for receiving complaints? Is it easy for individuals experiencing issues with the FRT system to find and use the complaint mechanism?

Note that some of the issues in this list cut across most if not all use cases, while others depend on the particular use case.

APPLYING THE FRAMEWORK TO REAL-WORLD USE CASES

The framework outlined in the preceding section is intended to identify issues that arise from the use of FRT in specific contexts. This section provides some examples of how the questions delineated in the risk management framework may provide helpful insight in concrete use cases. This section therefore applies portions of the risk management framework to four of the use cases introduced in Chapter 3—employee access control, aircraft boarding, protest surveillance, and retail loss prevention—to illustrate how the general questions posed in the framework play out in the context of specific uses and to develop a set of potential best practices for each case. These illustrative applications are brief and certainly do not consider every element of the risk framework, but they are intended to illustrate how a risk framework such as that suggested above can draw attention, in particular use cases, to key design and use issues that may enhance or detract from important values, like privacy and transparency. Encouraging (or requiring) that a framework be used to assess any given FRT invites organizations to, in essence, "show their work" and thus enhances transparency and, in many instances, can lead to greater care in system design.

Use of Facial Recognition Technology for Employee Access Control

Applying the risk management framework to the use of FRT for employee access control suggests that the following considerations—with respect to image collection, use, and retention; disclosure and consent; and fallback or alternative procedures—are of particular importance.

Image Collection, Use, and Retention

- Ensure that probe image collection is limited to select check-in locations such as a building entrance or security checkpoint. This helps guarantee that images are only collected when operationally necessary—that is, when an employee presents themselves for access to the facility.
- Ensure that probe image retention periods are strictly limited. For instances of controlling access to a facility, there is less need to keep the image for a long period of time. If, during the retention period, a probe image needs to be accessed and checked again (e.g., in case the employer wishes to determine whether a person was incorrectly granted access), administrators should seek organizational approval to access the image, documenting a specific purpose for which the image is needed.
- If an organization must share a probe image with another organization such as law enforcement, share only relevant probe images when data are requested and ensure that recipients also have adequate safeguards in place to limit the retention and use of images.
- Collect reference images when employees are hired and periodically update them in response to changes to the face from aging and technical needs for new systems.
- Store reference images in a secured system for managing access control and do not distribute or store them externally.
- Purge retained images after a set period of time when an employee leaves the organization or when a new reference image is collected.

Disclosure and Consent

- Ensure that cameras used to collect probe images are highly visible and feature signage detailing the purpose of the use of FRT and how captured images are used and retained.
- Organizations can assume consent from their employees and make enrollment mandatory given the legitimate business purpose of regulating access to the workplace but bear the responsibility for protecting reference images from disclosure.

Fallback or Alternative Procedures

- Use manual identification as a failsafe if an FRT system fails to verify the identity of an employee so that the employee is not incorrectly denied access.
- Use manual identification to regulate access to authorized visitors and non-employees from whom the organization may not have gained implied consent as a condition of employment.

Use of Facial Recognition Technology for Aircraft Boarding

Applying the risk management framework to the use of FRT as an alternative to other methods of identity verification when boarding an aircraft suggests that the following considerations—with respect to image collection, use, and retention; disclosure and consent; and fallback procedures—are of particular importance.

Image Collection, Use, and Retention

- Point equipment capturing probe images away from areas where passengers congregate to prevent the inadvertent photographing of any passenger who chooses to opt out of facial recognition.
- Retain reference images for limited time periods as established by local or federal regulations. Note that a long-term record of a passenger's identity is kept regardless of whether a passenger presents a boarding pass or uses facial recognition. However, associating an individual's identity with a flight does not require the long-term storage of biometric data.
- Require administrative approval and documentation if these data are to be kept for an extended period of time or shared with a third party.
- Share only relevant probe images when data are requested by law enforcement investigators and ensure that recipients also have adequate safeguards in place to limit the retention and use of images.
- Use the reference gallery of passengers[3] included in the manifest only for the purpose of boarding an aircraft and terminating access to the gallery once the aircraft has departed (unless needed by an international entity receiving the passengers).

Disclosure and Consent

- Ensure that cameras used to collect probe images are highly visible and feature signage detailing the purpose of the use of FRT and how captured images are used and retained.

---

[3] The reference gallery is collected from the Customs and Border Protection's Traveler Verification Service.

- Notify passengers of their right to opt out of facial recognition screening and establish alternate procedures to ensure that those opting out are not significantly delayed or inconvenienced.

Fallback Procedures

- Maintain existing procedures for verifying a passenger's claim to board an aircraft—for example, the ability to scan boarding passes and check physical documents—for passengers who choose to opt out of FRT identification.

Equity

- Collect statistics on whether members of particular demographic groups experience different FN match rates—that is, instances where individuals must physically present identification—and report the resulting aggregate time and inconvenience burdens.

Use of Facial Recognition Technology to Surveil a Protest

Applying the risk management framework to the use of FRT to surveil a protest suggests that the following considerations—with respect to image collection, use, and retention and disclosure and consent—are of particular importance.

Image Collection, Use, and Retention

- Strictly limit law enforcement image collection to defined public safety purposes so as to avoid a chilling effect on First Amendment rights.
- Use FRT only to identify individuals suspected of engaging in criminal behavior.
- Ensure that probe image retention periods are strictly limited to the time reasonably needed to conclude any criminal investigations that arise from an event.

Disclosure and Consent

- Develop and make publicly available policies that define the specific circumstances under which images are collected at public protests or submitted for FRT matching.

Use of Facial Recognition Technology to Assist in Retail Loss Prevention

The use of FRT for retail loss prevention differs from the use cases above because it takes place in a context where video surveillance has been widely used for decades. Applying the risk management framework to this use case suggests that the following considerations—with respect to image collection, use, and retention; disclosure and consent; and verification of an FRT match—are of particular importance:

Image Collection, Use, and Retention

- Include in the reference gallery of known shoplifters only individuals arrested for relevant offenses that were committed only in nearby geographic locations and within a set period of time.
- Before sharing a face image of a shoplifter known to one retailer with other retailers, consider whether the consequence of exclusion from multiple stores is warranted by the shoplifting threat the individual poses.

Disclosure and Consent

- Post prominent signs indicating that video surveillance and FRT are being used to identify known shoplifters and describe store procedures for handling customers identified using FRT as known shoplifters.

Verification of a Facial Recognition Technology Match

- If FRT identifies a customer as a known shoplifter, before taking action to remove the customer, dispatch a security guard or other store employee to obtain a government-issued photo identification from the customer and verify that the FRT identification was correct.

USE OF FACIAL RECOGNITION FOR LAW ENFORCEMENT INVESTIGATIONS

Applying the risk management framework to the use of FRT in law enforcement investigations suggests that it is important that (1) only validated (or certified, if a certification regime is established) FRT systems are used by law enforcement; (2) there is adequate training of users; (3) potential uses are defined and disclosed; (4) there is appropriate disclosure to an individual when FRT is at least one of the factors that has been used to identify them; (5) there are appropriate limits on law enforcement use that balance citizen privacy protections with public safety needs; and (6) there is adequate consideration given to the potential for disproportionate impacts on marginalized communities.

The committee offers the following recommendations to assist with the development of guidelines for responsible use of FRT by law enforcement and for law enforcement recipients of federal funding for FRT system deployment.

RECOMMENDATION 1-3: The Department of Justice and the Department of Homeland Security should establish a multi-disciplinary and multi-stakeholder working group on facial recognition technology (FRT) to develop and periodically review standards for reasonable and equitable

use, as well as other needed guidelines and requirements for the responsible use of FRT by federal, state, and local law enforcement. That body, which should include members from law enforcement, law enforcement associations, advocacy and other civil society groups, technical experts, and legal scholars, should be charged with developing

a. Standards for appropriate, equitable, and fair use of FRT by law enforcement.

b. Minimum technical requirements for FRT procured by law enforcement agencies and a process for periodically reevaluating and updating such standards.

c. Standards for minimum image quality for probe images, below which an image should not be submitted to an FRT system because of low confidence in any ensuing match. Such standards would need to take into account such factors as the type of investigation (including the severity of the crime and whether other evidence is available) and the resources available to the agency undertaking the investigation.

d. Guidance for whether FRT systems should (1) provide additional information about confidence levels for candidates or (2) present only an unranked list of candidates above an established minimum similarity score.

e. Requirements for the training and certification of law enforcement officers and staff and certification of law enforcement agencies using FRT as well as requirements for documentation and auditing. An appropriate body to audit this training and certification should also be identified.

f. Policies and procedures to address law enforcement failures to adhere to procedures or failure to attain appropriate certification.

g. Mechanisms for redress by individuals harmed by FRT misuse or abuse, including both damages or other remedies for individuals and mechanisms to correct systematic errors.

h. Policies for the use of FRT for real-time police surveillance of public areas so as to not infringe on the right of assembly or to discourage legitimate political discourse in public places, at political gatherings, and in places where personally sensitive information can be gathered such as schools, places of worship, and health care facilities.

i. Retention and auditing requirements for search queries and results to allow for proper oversight of FRT use.

j. Guidelines for public consultation and community oversight of law enforcement FRT.

k. Guidelines and best practices for assessing public perceptions of legitimacy and trust in law enforcement use of FRT.

l. Policies and standardized procedures for reporting of statistics on the use of FRT in law enforcement, such as the number of searches and the number of arrests resulting from the use of FRT, to ensure greater transparency.

RECOMMENDATION 1-4: Federal grants and other types of support for state and local law enforcement use of facial recognition technology (FRT) should require that recipients adhere to the following technical, procedural, and disclosure requirements:

a. Provide verified results with respect to accuracy and performance across demographics from the National Institute of Standards and Technology's Facial Recognition Technology Evaluation or similar government-validated third-party test.

b. Comply with the industry standards called for in Recommendation 1-2—or comply with future certification requirements, where certification would be granted on the basis of an independent third-party audit.

c. Use FRT systems that present only candidates who meet a minimum similarity threshold (and return zero matches if no candidates meet the threshold) rather than returning a fixed-length candidate list or "most-probable candidate" list when the output of an FRT system is being used for further investigation.

d. Adopt minimum standards for the quality of both probe and reference gallery images.

e. Use FRT systems only with a human-in-the-loop and not for automated detection of offenses, including issuing citations.

f. Limit the use of FRT to being one component of developing investigative leads. Given current technological capabilities and limitations, in light of present variations in training and protocols, and to ensure accountability and adherence with legal standards, FRT should be only part of a multi-factor basis for an arrest or investigation, in line with current fact-sensitive determinations of probable cause and reasonable suspicion.

g. **Restrict operation of FRT systems to law enforcement organizations that have sufficient resources to properly deploy, operate, manage, and oversee them (an adequate certification requirement would presumably ensure that such resources were in place).**

h. **Adopt policies to disclose to criminal suspects, their lawyers, and judges on a timely basis the role played by FRT in law enforcement procedural actions such as lead identification, investigative detention, establishing probable cause, or arrest.**

i. **Disclose to suspects and their lawyers, on arrest and in any subsequent charging document, that FRT was used as an element of the investigation that led to the arrest and specify which FRT product was used.**

j. **Publicly report on a regular basis de-identified data about arrests that involve the use of matches reported by FRT. The reports should identify the FRT system used, describe the conditions of use, and provide statistics on the occurrences of positive matches, false positive matches, and non-matches.**

k. **Publicly report cumulatively on any instances where arrests made partly on the basis of FRT are found to have been erroneous.**

l. **Conduct periodic independent audits of the technical optimality of an FRT system and the skills of its users, determining whether its use is indeed cost-justified.**

Even if not subject to federal grant conditions, state and local agencies should adopt these standards.

RESEARCH AND DEVELOPMENT

Public research organizations such as NIST already undertake important work in setting benchmarks and evaluating the performance of FRT systems. Additional government support could help NIST answer important questions on the performance of FRT systems in non-cooperative settings, how to improve data sets to both preserve privacy and promote equity in the performance of FRT tools, and how best to continue recent work on characterizing, understanding, and mitigating phenotypical disparities.

RECOMMENDATION 1-6: The federal government should support research to improve the accuracy and minimize demographic biases and to further explore the sociotechnical dimensions of current and potential facial recognition technology uses.

To understand better how to responsibly deploy FRT while protecting equity, fairness, and privacy, NIST, the Department of Homeland Security's Maryland Test Facility, or a similarly well-suited institution should conduct research on

- The accuracy of FRT systems in a variety of non-optimal settings, including non-optimal facial angle, focus, illumination, and image resolution.
- The development of representative training data sets for template extraction and other methods that developers can safely apply to existing data sets and models to adjust for demographic mismatches between a given data set and the public.
- The performance of FRT with very large galleries (i.e., tens or hundreds of millions of entries) to better understand the impacts of FP and FN match rates as the size of galleries used continues to grow.

To advance the science of FRT and to better understand the sociotechnical implications of FRT use, the National Science Foundation or a similar research sponsor should support research on

- Developing privacy-preserving methods to prevent malicious actors from reverse-engineering face images from stored templates.
- Mitigating FP match rate variance across diverse populations and building better understanding of the levels at which residual disparities will not significantly affect real-world performance.
- Developing approaches that can reduce demographic and phenotypical disparities in accuracy.
- Developing accurate and fast methods for directly matching an encrypted probe image template to an encrypted template or gallery—for example, using fully homomorphic encryption.
- Developing robust methods to detect face images that have been deliberately altered by either physical means such as masks, makeup, and other types of alteration or by digital means such as computer-generated images.
- Determining whether FRT use deters people from using public services, particularly members of marginalized communities.

- Determining how FRT is deployed in non-cooperative settings, public reaction to this deployment, and its impact on privacy.
- Determining how FRT may be used in the near future by individuals for abusive purposes, including domestic violence, harassment, political opposition research, etc.
- Determining how private actors might use FRT in ways that mimic government uses, such as homeowners who deploy FRT for private security reasons.
- Researching future uses of FRT, and their potential impacts on various subgroups of individuals.

BIAS AND TRUSTWORTHINESS

RECOMMENDATION 2: Developers and deployers of facial recognition technology should employ a risk management framework and take steps to identify and mitigate bias and cultivate greater community trust.

FRT has engendered mistrust about bias in its technological underpinnings and broader mistrust, especially in minority communities, about the role of technology in law enforcement and similar contexts.

RECOMMENDATION 2-1: Organizations deploying facial recognition technology (FRT) should adopt and implement a risk management framework addressing performance, equity, privacy, civil liberties, and effective governance to assist with decision making about appropriate use of FRT.

Until the recommended risk management framework is developed, the issues listed in Recommendation 1-5 may serve as a useful point of departure.

RECOMMENDATION 2-2: Institutions developing or deploying facial recognition technology should take steps to identify and mitigate bias and cultivate greater community trust—with particular attention to minority and other historically disadvantaged communities. These should include

a. **Adopting more inclusive design, research, and development practices.**

b. **Creating decision-making processes and governance structures that ensure greater community involvement.**
   c. **Engaging with communities to help individuals understand the technology's capabilities, limitations, and risks.**
   d. **Collecting data on false positive and false negative match rates in order to detect and mitigate higher rates found to be associated with particular demographic groups.**

Such practices will help address mistrust about bias in FRT's technological underpinnings and broader mistrust, especially in minority communities, about the role of technology in law enforcement and similar contexts.

POTENTIAL EXECUTIVE ACTION AND LEGISLATION

An outright ban on all FRT under any condition is not practically achievable, may not necessarily be desirable to all, and is in any event an implausible policy, but restrictions or other regulations are appropriate for particular use cases and contexts.

Concerns about the impacts of FRT intersect with wider questions about how to protect consumer privacy, where and how to limit government surveillance that could infringe on civil liberties, and more generally how to govern and regulate a proliferation of artificial intelligence and other powerful computing technologies.

> **RECOMMENDATION 3: The Executive Office of the President should consider issuing an executive order on the development of guidelines for the appropriate use of facial recognition technology by federal departments and agencies and addressing equity concerns and the protection of privacy and civil liberties.**

Comprehensively addressing such questions, especially to address nongovernmental uses, may require new federal legislation.

> **RECOMMENDATION 4: New legislation should be considered to address equity, privacy, and civil liberties concerns raised by facial recognition technology, to limit harms to individual rights by both private and public actors, and to protect against its misuse.**

Legislation should consider

a. *Limitations on the storing of face images and templates*. Legislation could, for example, prohibit the storing of face images or templates in a gallery unless the gallery will be used for a specifically allowed purpose. Inclusion in a gallery might, for example, be prohibited except under the following conditions:
   - *For prescribed government functions*, such as at the border or at international arrival and departure points to identify persons entering and leaving the country, using photos from government databases.
   - *Where there is explicit consent for a specific purpose*, such as a person setting up a new smartphone consenting to using FRT to unlock the phone or a person explicitly consenting to an airline's use of their passport photo to enable the person to check in and board flights using FRT.
   - *Where there are threats to life and physical safety*, such as by a performance venue to scan for specific individuals who have been reported by police as posing a threat to the life or physical safety of a performer or by a shelter for abuse victims to scan people arriving at the facility to find individuals subject to restraining orders prohibiting their interaction with residents of the shelter.

   An additional set of issues with respect to inclusion in galleries relates to collection and use of images gathered from websites and social media platforms—both whether it is appropriate to use these without consent or knowledge as well as the implications of including low-quality or synthetic images collected in this manner. Under current law, the fact that a gallery was created by harvesting facial images from the Web in violation of platforms' terms of service does not create a barrier to the instrument's usage. Of course, Congress, a state legislature, or even a policing authority could promulgate a new rule barring the use of FRT applications developed without the benefit of consent from those whose data is used for training purposes.

   Precisely which uses are or are not allowed merits careful consideration by legislators and the public at large. The risk management framework discussed earlier may provide a useful tool for considering these questions.

b. *Specific uses of concern*. Such uses might, for example, include the following:
   - Commercial practices that implicate privacy (through either broader privacy legislation addressing FRT risks or an FRT-specific federal privacy law);
   - Harassment or blackmail;

- Unwarranted exclusion from public or quasi-public places;
- Especially sensitive government FRT uses (e.g., pertaining to law enforcement or access to public benefits or federally subsidized housing);
- Public and private uses that tend to chill the exercise of political and civil liberties—both intentional and from the emergent properties of use at scale; and
- Mass surveillance or individual surveillance other than that properly authorized for law enforcement or national security purposes.

c. *User training.* In applications where the operator or other user is expected to apply judgment or discretion in when or how to use FRT systems or in interpreting their results, and where a false match may result in significant consequences for an individual, legislation could require training for the operators and decision-makers. A notable example of this type of application is law enforcement investigations. By contrast, there are applications where the fallback in case of a failure is simply to inspect a government-issued identity document; training may be less critical for such use cases.

d. *Certification.* Legislation could require certification of operators and other users and/or certification of organizations that operate FRT systems for applications where technical or procedural errors can significantly harm subjects, notably in law enforcement.

In light of the fact that FRT has the potential for mass surveillance of the population, courts and legislatures will need to consider the implications for constitutional protections related to surveillance, such as due process and search and seizure thresholds and free speech and assembly rights.

In grappling with these issues, courts and legislatures will have to consider such factors as who uses FRT, where it is used, what is it being used for, under what circumstances it is appropriate to use FRT-derived information provided by third parties, whether its use is based on individualized suspicion, intended and unintended consequences, and susceptibility to abuse, while courts will have to determine how constitutional guarantees around due process, privacy, and civil liberties apply the deployment of FRT.

As governments and other institutions take affirmative steps through both law and policy to ensure the responsible use of FRT, they will need to take into account the views of government oversight bodies, civil society organizations, and affected communities to develop appropriate safeguards.

Appendixes

A

Statement of Task

A National Academies of Sciences, Engineering, and Medicine study will assess current capabilities, future possibilities, societal implications, and governance of facial recognition (FR) technologies. It will

- Provide a broadly accessible explanation of FR technologies, their relationship to artificial intelligence and machine learning technologies, applications of FR technologies, and interactions and interoperability of FR technologies with other systems;
- Review existing governmental and other efforts aimed at explaining the workings and implications of FR technologies;
- Assess the strengths, capabilities, risks, and limitations of FR technologies, to include measures of performance and cost and differential accuracy across subpopulations (e.g., across races, genders, and ages);
- Consider current approaches to governing the use of FR technologies in law enforcement, non–law enforcement, and other common use cases and describe implications of the use of FR technology and requirements for adequate safeguards;
- Consider concerns about the impacts of FR technologies in public and private settings on privacy, civil liberties, and human rights, including issues of usability, equity, fairness, privacy, consent, community interests, and other societal considerations affecting FR acceptability; and

- Develop recommendations to govern the use and performance of facial recognition technologies in ways that could increase quality and efficiency, increase public safety, and safeguard privacy, civil liberties, and human rights.

B

Presentations to the Committee

JULY 7, 2022

> John Boyd, Office of Biometric Identity Management, Department of Homeland Security (DHS)
>
> Patrick Grother, Information Technology Laboratory, National Institute of Standards and Technology (NIST)
>
> Kate McKenzie, National Security and Law Enforcement Unit, Department of Justice
>
> John Howard, Maryland Test Facility
>
> Jody Hardin, Office of Field Operations, Customs and Border Protection
>
> Arun Vemery, DHS
>
> Benji Hutchinson, Paravision

JULY 8, 2022

> Clare Garvie, Center on Privacy and Technology, Georgetown University

SEPTEMBER 9, 2022

Federal Identity Forum and Expo

Amy Yates, NIST

Karl Ricanek, University of North Carolina Wilmington

Arun Ross, Michigan State University

SEPTEMBER 27, 2022

Daniel Heltemes, Forensic Images Unit, Arizona Department of Public Safety

Johanna Morley, INTERPOL; former Metropolitan Police (London)

David Russell, NOVARIS, Fairfax County (Virginia) Police

Jason Lim, Program Analyst, Transportation Security Administration

Meenakshi Nieto, Hartfield-Jackson Atlanta International Airport

Daniel Tanciar, Pangiam

Richard W. Vorder Bruegge, Federal Bureau of Investigation

OCTOBER 18, 2022

Daniel Bachenheimer, Accenture

Brenda Leong, BNH.ai

Jennifer Lynch, Electronic Frontier Foundation

Jay Stanley, American Civil Liberties Union

OCTOBER 31, 2022

James Wayman, IET Biometrics

NOVEMBER 18, 2022

Safiya Noble, Center for Critical Internet Inquiry, University of California, Los Angeles

Aylin Caliskan, University of Washington

Michael Akinwumi, National Fair Housing Alliance

Brenda Goss Andrews, National Organization Black Law Enforcement Executives

Bertram Lee, Data, Decision Making, and Artificial Intelligence, Future of Privacy Forum

Rashawn Ray, Brookings Institution; University of Maryland

Tawana Petty, Algorithmic Justice League

Renee Cummings, University of Virginia

Paromita Shah, Just Futures Law

DECEMBER 15, 2022

Representative Ted Lieu (D-CA), U.S. House of Representatives

DECEMBER 16, 2022

Michael Kearns, University of Pennsylvania

JANUARY 31, 2023

Stephane Gentric, IDEMIA

APRIL 14, 2023

John Mears, Homeland Security Solutions, Leidos

Donnie Scott, IDEMIA Identity and Security

Neville Pattinson, Thales

C

Committee Member Biographical Information

EDWARD W. FELTEN, *Co-Chair*, is the Robert E. Kahn Professor of Computer Science and Public Affairs, Emeritus, at Princeton University and was the founding director of Princeton's Center for Information Technology Policy, a cross-disciplinary effort studying digital technologies in public life. Dr. Felten is also the co-founder and chief scientist at Offchain Labs, Inc., and he is a member of the U.S. Privacy and Civil Liberties Oversight Board. From 2015 to 2017, Dr. Felten served on the White House staff as Deputy United States Chief Technology Officer. His research interests include blockchain and cryptocurrency technologies, computer security and privacy, and technology law and policy. He has published more than 150 papers in the research literature and 2 books. His research on topics such as Web security, copyright and copy protection, and electronic voting has been covered extensively in the popular press. Dr. Felten is a member of the American Academy of Arts & Sciences and a fellow of the Association for Computing Machinery (ACM). He has testified at House and Senate committee hearings on privacy, electronic voting, digital television, and artificial intelligence (AI). Among his committee service at the National Academies of Sciences, Engineering, and Medicine, Dr. Felten served on the Committee on Directions for the AFOSR Mathematics and Space Sciences Directorate Related to Information Science and Technology (2005) and the Committee on the Fundamentals of Computer Science: Challenges and Opportunities (2000–2005). He earned a PhD in computer science and engineering from the University of Washington. Dr. Felten served in an uncompensated advisory role for the Center for Democracy and Technology from 2017 to 2019.

JENNIFER L. MNOOKIN, *Co-Chair*, is the chancellor of the University of Wisconsin–Madison. Before that, she was Ralph and Shirley Shapiro Professor of Law, a member of the University of California, Los Angeles (UCLA), School of Law faculty, and the dean of the UCLA School of Law. Dr. Mnookin is a leading evidence scholar and among the most highly cited law faculty in her field. She has been a co-author of two major evidence treatises, "The New Wigmore, A Treatise on Evidence: Expert Evidence" and "Modern Scientific Evidence: The Law and Science of Expert Testimony" and has published extensively on issues relating to evidence, expertise, and the use of specialized knowledge in the lay adjudicatory system. In 2020, Dr. Mnookin was elected to the American Academy of Arts & Sciences. She served for 6 years on the National Academies of Sciences, Engineering, and Medicine's Committee on Science, Technology, and Law (2014–2020). She co-chaired a group of senior advisors for a President's Council of Advisors on Science and Technology report on the use of forensic science in criminal courts, and currently serves on the board of the Law School Admissions Council. Prior to joining the UCLA School of Law, Dr. Mnookin was a professor of law and the Barron F. Black Research Professor at the University of Virginia School of Law and a visiting professor of law at Harvard Law School. She received her AB from Harvard University, her JD from Yale Law School, and a PhD in history and social study of science and technology from the Massachusetts Institute of Technology. Dr. Mnookin also serves on the non-governing, uncompensated advisory board of the Electronic Privacy Information Center (EPIC).

THOMAS D. ALBRIGHT is a professor and the Conrad T. Prebys Chair at the Salk Institute for Biological Studies. Dr. Albright is an authority on the neural basis of visual perception, memory, and visually guided behavior. His laboratory seeks to understand how visual perception is affected by attention, behavioral goals, and memories of previous experiences. An important goal of this work is the development of therapies for blindness and perceptual impairments resulting from disease, trauma, or developmental disorders of the brain. A second aim of Dr. Albright's work is to use our growing knowledge of brain, perception, and memory to inform design in architecture and the arts, to leverage societal decisions and public policy, and to advise on matters of law and justice. He is a fellow of the American Academy of Arts & Sciences and a fellow of the American Association for the Advancement of Science. Dr. Albright served on the National Academies' Committee on Science, Technology, and Law and on the National Commission on Forensic Science, an advisory body to the Department of Justice. Dr. Albright is currently a member of the Human Factors Resource Committee of the National Institute of Standards and Technology's (NIST's) Organization of Scientific Area Committees for Forensic Science. He served as the co-chair of the National Academies' Committee on Scientific Approaches to Eyewitness Identification, which produced the 2014 consensus study report *Identifying the Culprit: Assessing Eyewitness Identification*. Dr. Albright received a PhD in psychology and neuroscience from Princeton University.

RICARDO BAEZA-YATES is the director of research at the Institute for Experiential Artificial Intelligence at Northeastern University. Dr. Baeza-Yates came to Northeastern after his role as the chief technology officer of NTENT, a semantic search technology company based in California. Prior to these roles, he was the vice president of research at Yahoo Labs, based in Sunnyvale, California, from 2014 to 2016. Before joining Yahoo Labs in California, he founded and led the Yahoo Labs in Barcelona, Spain, and Santiago, Chile, from 2006 to 2015. Between 2008 and 2012, he oversaw Yahoo Labs in Haifa, Israel, and started the London laboratory in 2012. Dr. Baeza-Yates is a part-time professor in the Department of Information and Communication Technologies at the Universitat Pompeu Fabra in Barcelona, and in the Department of Computing Science at the Universidad de Chile in Santiago. In 2005, he was an ICREA research professor at the Universitat Pompeu Fabra. Until 2004, he was a professor and the founding director of the Center for Web Research at the Universidad de Chile. Additionally, he is a co-author of the best-selling textbook *Modern Information Retrieval* (1999), with a second enlarged edition in 2011 that won the ASIST 2012 Book of the Year award. He is also a co-author of the second edition of the *Handbook of Algorithms and Data Structures* (1991) and a co-editor of *Information Retrieval: Algorithms and Data Structures* (1992), among more than 600 publications. From 2002 to 2004, he was elected to the board of governors of the Institute of Electrical and Electronics Engineers (IEEE) Computer Society, as well as to the ACM Council from 2012 to 2016. Dr. Baeza-Yates has received the Organization of American States award for young researchers in exact sciences, the Graham Medal for innovation in computing given by the University of Waterloo, Canada, to distinguished alumni, the CLEI Latin American distinction for contributions to computer science in the region, and the National Award of the Chilean Association of Engineers, among other distinctions. In 2003, he was the first computer scientist to be elected to the Chilean Academy of Sciences and, since 2010, is a founding member of the Chilean Academy of Engineering. In 2009, he was named an ACM fellow and, in 2011, an IEEE fellow. Dr. Baeza-Yates received his PhD in computer science from the University of Waterloo. He served on the ACM Technology Policy Council when it issued a June 2020 statement urging the suspension of the use of facial recognition systems.

ROBERT BLAKLEY is an operating partner at Team8. He was previously the global director of information security innovation at Citi. Dr. Blakley co-hosts the IEEE podcast "Over the Rainbow: 21st Century Security and Privacy." He recently served as a member of the National Academies' Committee on Technical Assessment of the Feasibility and

Implications of Quantum Computing and as a member of the National Academies' Forum on Cyber Resilience. Dr. Blakley has served as the plenary chair of the National Strategy for Trusted Identities in Cyberspace Identity Ecosystem Steering Group and as the research and development co-chair of the Financial Services Sector Coordinating Council for Critical Infrastructure Protection and Homeland Security and was the general editor of the OASIS SAML specification. Prior to joining Citi, Dr. Blakley was the distinguished analyst and agenda manager for identity and privacy at Gartner and Burton Group. Before that, he was the chief scientist for security and privacy at IBM. He is the past general chair of the IEEE Security and Privacy Symposium and the Applied Computer Security Associates New Security Paradigms workshop. He was awarded the Annual Computer Security Applications Conference's Distinguished Security Practitioner award in 2002 and is a frequent speaker at information security and computer industry events. Dr. Blakley received an AB in classics from Princeton University and an MS and a PhD in computer and communications science from the University of Michigan.

PATRICK GROTHER is a scientist in the Information Technology Laboratory at NIST, a division of the Department of Commerce. Mr. Grother is responsible for biometric standards and technology evaluation at NIST and leads its Face Recognition Vendor Test program, the world's largest independent public testing program of face recognition algorithms. He co-chairs NIST's biannual International Face Performance Conference on measurement, metrics, and certification and assists several U.S. government agencies in biometrics performance assessment and standardization. Since 2018, Mr. Grother has served as the chair of the ISO/IEC/JTC 1 Subcommittee 37 on Biometrics, where he is the editor of five international standards. He is a three-time recipient of the Department of Commerce Gold Medal award for service to U.S. industry. Mr. Grother has an MS in computer science from Imperial College London.

MARVIN B. HAIMAN serves as the chief of staff for the Metropolitan Police Department, Washington, DC. In this capacity, Mr. Haiman oversees daily operations of the Executive Office of the Chief of Police and is responsible for broad agency management and implementing strategic agency objectives. He is also responsible for several organizational units, including the Office of Communications, Office of General Counsel, and Professional Development Bureau. Mr. Haiman served as the executive director of the Professional Development Bureau from 2017 to 2021, leading recruiting, training, human resources, discipline, promotional processes, equal employment opportunity, and volunteer services functions. Prior to being named the executive director, from 2015 to 2017, Mr. Haiman served in a variety of capacities, including developing the agency's Office of Volunteer Coordination, serving as the chief of staff for the Technical Services

Division tasked with a broad range of information technology operations. Before returning to the Metropolitan Police Department, he served as the director for the Homeland Security Advisory Council for the Department of Homeland Security, establishing several key task forces for the Secretary. Mr. Haiman is also a graduate of the Naval Postgraduate School Center for Homeland Defense and the Security's Executive Leadership Program. He was recognized by the International Association of Chiefs of Police in 2020 as a 40 under 40 recipient and received the prestigious Gary P. Hayes Award from the Police Executive Research Forum. Mr. Haiman also graduated from the Metropolitan Police Academy and served as a reserve police officer with the Metropolitan Police Department. He graduated from Johns Hopkins University with a master's degree in management through the Police Executive Leadership Program, has an undergraduate degree in mathematics from The University of Iowa, and has earned certifications in public management and strategic project management from The George Washington University.

AZIZ Z. HUQ is a scholar of U.S. and comparative constitutional law at The University of Chicago. Mr. Huq works on topics ranging from democratic backsliding to regulating AI. His award-winning scholarly work is published in several books and in leading law reviews, social science journals, and political science journals. He has also written for *The Washington Post, The New York Times*, *Dissent*, *The Nation*, and many other non-specialist publications. In 2015, he received the Graduating Students Award for Teaching Excellence. Mr. Huq has an active pro bono practice and is on the board of the American Constitution Society, the New Press, and the American Civil Liberties Union (ACLU) of Illinois. Before joining the Law School faculty, he worked as the counsel and then director of the Brennan Center's Liberty and National Security Project, litigating cases in both the U.S. Court of Appeals and the Supreme Court. As a senior consultant analyst for the International Crisis Group, he researched and wrote on constitutional design and implementation in Pakistan, Nepal, Afghanistan, and Sri Lanka. Mr. Huq was a law clerk for Judge Robert D. Sack of the U.S. Court of Appeals for the Second Circuit and then for Justice Ruth Bader Ginsburg of the Supreme Court of the United States. He is also a 1996 summa cum laude graduate of the University of North Carolina at Chapel Hill and a 2001 graduate with a JD from Columbia Law School, where he was awarded the John Ordronaux Prize (for the student graduating first in their class). He has served as an uncompensated board member of the ACLU of Illinois since 2018 and an uncompensated board member of the American Constitution Society since 2019. Mr. Huq's writings on issues relevant to facial recognition include "The Public Trust in Data," *Georgetown Law Journal* (2021); "Privacy's Political Economy and the State of Machine Learning: An Essay in Honor of Stephen J. Schulhofer," *NYU Annual Survey of American Law* (2021; co-authored with Mariano-Florentino Cuéllar); "Constitutional Rights in the Machine

Learning State," *Cornell Law Review* (2020); and "Racial Equity in Algorithmic Criminal Justice," in the *Duke Law Journal* (2019).

ANIL K. JAIN is a University Distinguished Professor at Michigan State University, where he conducts research in pattern recognition, computer vision, and biometrics recognition. He served as a member of the Defense Science Board and the Forensics Science Standards Board. Dr. Jain received Guggenheim, Humboldt, and Fulbright fellowships and the King-Sun Fu Prize. He was the editor-in-chief of *IEEE Transactions on Pattern Analysis and Machine Intelligence* and is a fellow of ACM and IEEE. For advancing pattern recognition and biometrics, Dr. Jain was awarded Doctor Honoris Causa by Universidad Autónoma de Madrid and Hong Kong University of Science and Technology. Dr. Jain received his PhD in electrical engineering from The Ohio State University.

ELIZABETH E. JOH is a professor of law at the University of California, Davis, School of Law. Dr. Joh researches and writes primarily in the areas of policing, surveillance, and technology. Her work in this area focuses on the impacts of new technologies on democratic policing and its effects on privacy and civil liberties. She has written in nationally prominent law journals and in publications including *The New York Times*, *Slate*, and the *Los Angeles Times*. Dr. Joh is an elected member of the American Law Institute, a faculty advisory board member of the University of California CITRIS Policy Lab, an affiliate scholar with Stanford University's Center for Internet and Society, and served on the University of California Presidential Working Group on Artificial Intelligence. Dr. Joh received her JD and PhD in law and society from New York University. She has served as an uncompensated advisory board member of EPIC since 2019. Dr. Joh's previous writings on issues related to the committee's work include "The Corporate Shadow in Public Policing," *Science* (2021); "Increasing Automation in Policing," *Communications of the ACM* (2020); and "Police Surveillance Machines: A Short History," *Law and Political Economy* (2018).

MICHAEL C. KING is an associate professor in the Department of Electrical Engineering and Computer Science and a research scientist for the L3-Harris Institute for Assured Information at the Florida Institute of Technology and has served in this role since 2015. Before joining academia, Dr. King served 14 years as a scientific research/program management professional in the U.S. Intelligence Community. While in government, Dr. King created and managed research portfolios covering a broad range of topics related to biometrics and identity. He crafted and successfully led the Intelligence Advanced Research Projects Activity's Biometric Exploitation Science and Technology Program to transition technology deliverables to more than 40 government organizations. As a subject-matter expert in biometrics and identity intelligence, Dr. King has been invited to brief the director of National Intelligence, congressional staffers and science advisers, the Defense Science Board, the Army Science Board, and the Intelligence Science Board. He also served as Intelligence Community Department Lead to the White House Office of Science and Technology Policy's National Science and Technology Council Subcommittee on Biometrics and Identity Management. Additionally, Dr. King worked for the National Security Agency as the technical director in the Research Directorate's Human Interface Security Research Focus Area. Dr. King's final government appointment was as the director of applied research and innovation in the Directorate of Science and Technology's Office of Technical Services at the Central Intelligence Agency. In this role, he was responsible for leading a team of scientists in the delivery of advanced capabilities in cyber, identity intelligence, and special communication systems. Dr. King received his PhD in electrical engineering from North Carolina Agricultural and Technical State University. Dr. King briefly consulted for Idemia-National Security Solutions from January to March 2020. He has spoken on facial recognition at Idemia user conferences in 2017 and 2019, for which his travel was compensated. He has received and evaluated software from Idemia, Rank One Computing, and Cyberextruder, for which he received no compensation.

NICOL TURNER LEE is a senior fellow in governance studies and the director of the Center for Technology Innovation at The Brookings Institution. She also serves as the co-editor-in-chief of *TechTank* and the co-host of the #TechTank podcast. Dr. Turner Lee's research explores public policies designed to enable equitable access to technology, especially those that create systemic changes in global communities. Her work focuses on global and domestic broadband deployment, AI, and various Internet governance concerns. Dr. Turner Lee is an expert on the intersection of race, technology, and social justice. She has a forthcoming book on the U.S. digital divide titled *Digitally Invisible: How the Internet Is Creating the New Underclass* (2024). She has also written various papers and book chapters on inclusive and equitable AI. Currently, she serves as a vice chair of the Federal Communications Commission's Communications Equity and Diversity Council. Dr. Turner Lee is a graduate of Colgate University magna cum laude and has an MA and a PhD in sociology from Northwestern University. She also holds a certificate in nonprofit management from the University of Illinois Chicago.

IRA REESE is currently the chief technology officer of the Washington, DC–based firm Global Security and Innovative Solutions and consults in the areas of border and sports venue security as well as specific issues regarding international trade. Mr. Reese is an internationally recognized expert in border security and the equipment and software used to secure borders. He was formerly the Senior Executive Service executive director of laboratories and scientific services of Customs and Border Protection for 16 years and responsible for the management of field laboratories, forensic units, an engineering branch, and the 24×7 Weapons of Mass Destruction (WMD) response center. These units have 350 scientists and engineers specializing in trade issues, development of border security protocols and related equipment procurement, narcotics and human trafficking enforcement, approval of U.S. international pipelines, and licensing of commercial gaugers. During this time, Mr. Reese was the six-times-elected chair (2-year terms) of the World Customs Organization Scientific Sub-Committee, representing 171 countries. He has received the Presidential Award from the American Health Physics Society for the development, deployment, and maintenance of the U.S. Radiation WMD Ports Security Program and is a graduate of the Harvard University Senior Executive Fellows Program. During his time in federal service, Mr. Reese carried Top Secret, Top Secret/SCI, and Q security clearances.

CYNTHIA RUDIN is a professor of computer science and engineering at Duke University and directs the Interpretable Machine Learning Lab. Her goal is to design predictive models that people can understand. Dr. Rudin is the recipient of the 2022 Squirrel AI Award for Artificial Intelligence for the Benefit of Humanity from the Association for the Advancement of Artificial Intelligence (AAAI) (the "Nobel Prize of AI"). She is also a three-time winner of the INFORMS Innovative Applications in Analytics Award and is a 2022 Guggenheim Fellow. Dr. Rudin is a fellow of the American Statistical Association, the Institute of Mathematical Statistics, and AAAI. She has served on three National Academies' committees, including the Committee on Applied and Theoretical Statistics, the Committee on Law and Justice, and the Committee on Analytic Research Foundations for the Next-Generation Electric Grid. Dr. Rudin received her PhD in applied and computational mathematics from Princeton University. In 2021, she submitted a letter to the White House on the use of AI in criminal justice and co-authored the article "A Truth Serum for Your Personal Perspective on Facial Recognition Software in Law Enforcement" in *Translational Criminology*.

D

Disclosure of Unavoidable Conflict of Interest

The conflict-of-interest policy of the National Academies of Sciences, Engineering, and Medicine (https://www.nationalacademies.org/coi) prohibits the appointment of an individual to a committee like the one that authored this Consensus Study Report if the individual has a conflict of interest that is relevant to the task to be performed. An exception to this prohibition is permitted only if the National Academies determine that the conflict is unavoidable and the conflict is promptly and publicly disclosed.

Professor Anil Jain has a conflict of interest in relation to his service on the Committee on Facial Recognition: Current Capabilities, Future Prospects, and Governance because he is a consultant for Amazon on technologies relating to palm print recognition and has stock holdings in Microsoft Corporation, Amazon.com, Inc., and Rank One Computing.

The National Academies have concluded that for this committee to accomplish the tasks for which it was established, its membership must include at least one person who has current experience in the development and application in industry of facial recognition technologies, biometrics, pattern recognition, and computer vision. As described in his biographical summary, Professor Jain has current experience working on design and applications of pattern recognition systems, focusing on automatic fingerprint recognition, automatic face recognition, and large-scale data clustering.

The National Academies have determined that the experience and expertise of Professor Jain are needed for the committee to accomplish the task for which it has been established. The National Academies could not find another available individual with the equivalent experience and expertise who did not have a conflict of interest. Therefore, the National Academies concluded that the conflict was unavoidable and publicly disclosed it on its website (www.nationalacademies.org).